Energy Conservation through Control

ENERGY SCIENCE AND ENGINEERING:
RESOURCES, TECHNOLOGY, MANAGEMENT
An International Series
EDITOR
JESSE DENTON
Belton, Texas

ENERGY CONSERVATION THROUGH CONTROL

Francis G. Shinskey

Systems Development
The Foxboro Company
Foxboro, Massachusetts

ACADEMIC PRESS New York San Francisco London 1978
A Subsidiary of Harcourt Brace Jovanovich, Publishers

ACADEMIC PRESS, INC.
111 Fifth Avenue, New York, New York 10003

United Kingdom Edition published by
ACADEMIC PRESS, INC. (LONDON) LTD.
24/28 Oval Road, London NW1 7DX

Library of Congress Cataloging in Publication Data

Shinskey, F   Greg.
    Energy conservation through control.

    (Energy series)
    Includes bibliographical references.
    1. Energy conservation.   2.   Process control.
I.   Title.
TJ163.3.S54   621   77-6613
ISBN 0-12-641650-8

PRINTED IN THE UNITED STATES OF AMERICA

80 81 82   9 8 7 6 5 4

Contents

Preface

Historically, control technology has focused its attention on minimizing the deviation of a variable from some set point arbitrarily given by others. While this pursuit has occupied many engineers and scientists for sizeable portions of their careers, it represents only part of the problems faced in the successful operation of a plant. Oddly enough, the practicing control engineer has far more influence over plant operations than his job responsibility would seem to allow. In fact, a proficient control engineer can squeeze more production out of limited plant equipment and can contribute more to reducing operating costs than almost any other individual in the plant.

If control technology is so restrictive, then how can a control engineer contribute so much to plant operation? The answer lies in the means that he must take to achieve his goal of effective control over important plant variables. If the controls are applied with little understanding of how the plant functions or what it needs, they will perform poorly, and the operators will not use them. The most effective control system is an embodiment of the characteristics of the process being controlled, including its failings and limitations, its interactions and demands. In a good marriage, the plant and its controls will work as one, to the delight and satisfaction of manager and operator alike. But this is not easily achieved. To design the control system that effectively matches the plant requires an understanding of the plant rivaling that of the plant's designers, operators, and manager.

It is then not surprising that a control engineer may know more about the plant than perhaps any other individual associated with it. In such a role he may be called upon frequently to advise the plant manager on how to increase productivity or reduce costs—even to the point of modifying process equipment or altering operating procedures. In this guise, the control engineer has expanded his practice beyond the limits of what is commonly accepted as control technology—through his understanding of the process being controlled.

In this book, the author expands his horizons to include subjects not commonly in the domain of process control. The "systems" concept is used throughout, encompassing an ever-broadening sphere of influence. As an illustration, control of main steam pressure in a power plant is not restricted to manipulation of the turbine throttle valve or fuel to the burners. Such a narrow point of view contributed measurably to our impending energy shortfall. The characteristics of the turbine must be considered from a thermodynamic standpoint—the throttle valve robs the steam of much of its available work. Boiler efficiency must also be examined, along with the availability of waste fuels and energy impounded in storage. Consideration must also be given to balancing the demand for shaft work against heat usage to maximize the utilization of the energy released by combustion. While the first consequence of this sort of perspective will be the implementation of energy-conserving control systems, ultimately it should have an influence on plant design and even contractual relationships between buyer and seller.

Processes involving energy conversion are placed at the forefront of this book. But they cannot be intelligently examined without an introduction to the oft-ignored and rarely understood "second law of thermodynamics." Were it not for a reassessment of the relationships between energy and work in light of the second law, the author would have comparatively little to contribute to "Energy Conservation through Control". But by equating irreversibility and loss in available work, the second law casts light on customs and practices that by their nature waste energy. After digesting Chapter 1, the reader should have little difficulty recognizing those customs and practices, and becoming motivated toward avoiding, eliminating, or minimizing their continuation. The author's recommendations in the areas of combustion control and particularly steam plant management are a natural consequence of applying these principles. Long established traditions such as constant-pressure operations and desuperheating by blending must be replaced by control methods that are equally effective in protecting equipment and maintaining production without the waste normally associated with control. The benefits are still more pronounced when these concepts are applied to the control of rotating machinery in the subsequent chapters on compressors and refrigeration units.

A large section of the book is devoted to mass-transfer operations because much, if not most, of the energy consumed in process plants and refineries is allocated to separation and purification. Here, the second law again applies, in that the energy content of raw feedstocks and refined products is often equal. The principal value of the products then lies in their purity, which requires work to achieve. Purification or separation then has a work-equivalent, and product-quality control is directly convertible into energy utilization.

The last part of the book addresses energy use in buildings. A large portion of the nation's energy is consumed in providing heat, ventilation, air-conditioning, and domestic hot water. It does not require a degree in architecture or civil engineering to realize how poorly planned and organized these systems are in modern office buildings. Heating and cooling, lighting and shading, are generally conducted *concurrently* in the same space. The result is not only waste but even discomfort for the occupants. Building design desperately needs a complete reevaluation—greater use of atmospheric cooling and natural lighting are essential to comfort and minimized dependence on fuels. Solar energy can play a major role in supplying building needs, and though in its infancy, the technology for its utilization is advancing rapidly, with production of usable components already beginning. No present-day book on energy would be complete without including this available, nonpolluting, timeless resource.

Let me express my sincere appreciation for the help of Marilyn Clark who typed the manuscript and Leo Paananen for his artwork.

Symbols

A	Airflow, area	M	Mass flow, molar flow
B	Bottom product flow	N	Speed
C	Specific heat, flow coefficient, correction factor	P	Proportional band
		Q	Heat, heat flow
D	Distillate flow	R	Gas constant, rate coefficient
E	Efficiency	S	Entropy
F	Flow, feed rate	T	Temperature
G	Gas specific gravity, gain, mass flow of gas	U	Heat-transfer coefficient
		V	Volume, volumetric flow, vapor
H	Enthalpy	W	Work, power
I	Integral time	X	Fractional flow
K	Constant	Z	Supercompressibility factor
L	Length, reflux flow	$\$$	Cost
a	Valve opening, constant	p	Pressure
b	Constant, bias	q	Load
c	Cost, constant	r	Rangeability, ratio
d	Differential	s	Separation factor
e	Deviation, 2.718	t	Time
f	Function, constant, feedback	u	Velocity
g	Constant, gravitational acceleration	v	Volume, specific volume, value
h	Head, differential pressure	w	Molecular weight, weight fraction
k	Constant	x	Mol fraction
m	Manipulated variable, valve position	y	Mol fraction
n	Number	z	Mol fraction

α	Relative volatility, constant	μ	Joule—Thompson coefficient
β	Column characterization factor	π	3.1416
γ	Wobbe Index, ratio of specific heats	ρ	Density
Δ	Difference	Σ	Sum
∂	Partial differential	τ	Time constant
η	Efficiency	ϕ	Function of specific-heat ratio
θ	Angle	χ	Variation

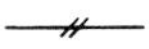

Control signal (pneumatic)

Control signal (electric)

Control signal (capillary)

Control valve

Control damper

Control louvers

Control valve with positioner

Solenoid or on–off valve

Nozzle or venturi

Magnetic flowmeter

Turbine flowmeter

Multiplier

Divider

Three-way valve
(normally closed port)

Controller, locally mounted
(liquid level)

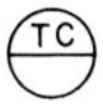

Controller, board-mounted
(temperature)

Controller, ratio, board-mounted
(flow)

Switch (pressure)

Transmitter (differential pressure)

Alarm unit

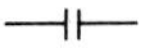

Orifice

High selector

Function generator

Dynamic characterizer

Bias station

Ratio station

Low limit

Symbols

	Square-root extractor		High limit
	Summer (adding and subtracting)		Integrator
	Low selector		Time-delay relay

Part

I

INTRODUCTION

Chapter

1

Thermodynamics and Energy Conservation

The *first law of thermodynamics* tells us that in any isolated system energy is naturally conserved. The physicist J. P. Joule was able to heat water in an insulated vessel by stirring, by an electric current, by cooling a compressed gas, and by friction between iron blocks, on a quantitative basis. Thus he established that the energy released by hydraulic, pneumatic, electrical, and mechanical systems is all convertible into heat.

Since energy is never lost, how then can it be in short supply? The answer is that energy is not in short supply—in fact, it is quite abundant, although little of it is in a usable form. Consider the earth as an isolated system (which it is not, in that it absorbs radiation from the sun which it ultimately re-radiates into space). All of the fossil fuels stored on earth at any given time represents an inventory of high-level energy, which may be released by combustion. If it were all to burn at once, the net energy content of the (isolated) earth would not change—the environment would increase in temperature to a degree corresponding exactly to the release of energy by the fossil fuels. We would have spent all our stored high-level energy only to raise the temperature of our environment. Without losing any actual quantity of energy, we would nonetheless be "energy bankrupt."

Energy at a level equal to that of the environment is useless for doing work. On a cold winter day, air at a temperature of 90°F would be welcome, because we could heat our homes with it. Yet on a hot summer day, that same air would be useless, because of its equilibrium with the environment. The value of a particular form of energy is then determined by its level (temperature) relative to that of the environment. Chilled water at 40°F is useful for refrigeration, while hot water at 120°F is useful for washing; but mix them together and the blend has little value.

Then the term "energy conservation" needs to be qualified if it is to be meaningful to our national effort. We could strive for "fuel conservation," and this terminology may satisfy most of our criteria. Yet a common denominator relating all forms of energy—fuels, electricity, steam, solar radiation, etc., is desirable to quantify our conservation effort. This common denominator is thermodynamics. By applying *the second law of thermodynamics*, it is possible to evaluate the usefulness of any given form and level of energy. Moreover, the efficiency of the process consuming that energy may be determined, which can lead the way to its improvement.

While Joule was able to convert work quantitatively into heat, the reverse is comparatively difficult. The *second law of thermodynamics* states that heat cannot be completely converted into work in a cyclic process (i.e., some residual heat will remain unconverted). The reason for this limitation is that equilibrium can only be approached by the flow of heat and/or mass from a higher to a lower energy level. These relationships can best be appreciated by introducing the concepts of entropy and availability.

A.   CONSERVATION OF AVAILABLE WORK

Energy is useful in two forms: heat and work. Work is the superior form because it can be converted completely into heat, while the reverse is not possible. Only a portion of the heat energy in a fluid is convertible into work in a complete cycle, depending on:

(a)   the initial and final states of the fluid, and
(b)   the thermodynamic efficiency of the process.

A greater amount of work can be extracted from a given mass of a very hot fluid (e.g., steam) than from a cooler fluid even though its quantity is sufficient to contain the same net energy relative to its surroundings. Fuel gas can be used to drive a turbine or heat a home; while air at 120°F can heat a home its temperature is not high enough to drive a turbine. Consequently, fuel gas should not be used to heat homes, but reserved for those processes which can make best use of its *available work*. Conversely, home heating can be accomplished using fluids containing little available work, such as solar-heated water or turbine-exhaust steam.

Fluids at elevated temperature and fuels yielding combustion products having elevated temperatures are then much more valuable than their enthalpy alone would indicate. A fair measure of their value is their available work, by which they can be compared to electrical or mechanical energy, which are directly convertible to work.

To illustrate the concept of conserving available work, consider a power plant fired with oil, sending electricity to a nearby community. Let the homes in the community also be heated with oil. By the nature of the fluid cycle in the power plant, only about 35% of the energy in its fuel is convertible to electricity, the rest being rejected to the surroundings. At the same time, the residences are rejecting a certain amount of energy to the surroundings as heat loss. If the waste heat from the power plant were used to heat homes in the community, an equivalent amount of fuel could be saved. But if the homes were treated electrically, a far worse situation would develop, in that the power plant must consume nearly three times as much fuel as the homes would need to satisfy their heating requirements.

The use of fuel for heating alone results in loss of its work content, which in the power plant cycle is 35% of its energy content. But the use of electricity for heating alone results in 100% loss in its work content. Electricity is more profitably used to drive motors, or to power electronic equipment. By weighing sources of energy on the basis of their available work, we may more efficiently allocate their use, and ultimately conserve energy.

The *processes* wherein energy is used or converted may also be evaluated as to how effectively they conserve work. All processes are subject to physical limitations such as friction. But some processes are theoretically very efficient while others are not. While the physical limitations can only be minimized by careful fabrication, etc., processes which are inherently inefficient are to some extent *avoidable*. In fact, many processes are purposely made inefficient to allow them to be controlled, as brakes are used to control a car. Some of these losses can be avoided altogether, and many more can be reduced. But an understanding of the concept of thermodynamic inefficiency is prerequisite to defining the problem.

1.   Available Work and Reversible Processes

A process in which the state of a fluid is changed by applied work, and which can produce the same amount of work by returning to its original state, is a *reversible* process. An example of a reversible process is an ideal compressor acting on an ideal gas. Work applied to the compressor can raise the pressure and temperature of the gas. If the system is perfect, i.e., without friction or heat loss, expansion of the gas to its original state can produce exactly the same amount of work as was used to compress it. While frictional and heat losses and fluid nonideality will, in practice, result in less work being recovered than applied, the compression–expansion process is at least *theoretically reversible.*

But if the compressed gas is instead expanded through an orifice and cooled to its original state, *no work is recovered.* The operations of expansion through an orifice and heat transfer are *theoretically irreversible.* After an introduction to the thermodynamic concepts relating heat and work, it will be possible to identify those processes which, by nature, are irreversible.

Consider the heat engine shown in Fig. 1.1. It is theoretically impossible to convert all of the energy in the hot fluid into work. But the *maximum* conversion for any given change in state will be achieved in a *reversible* process. The amount of heat that is rejected if the process is reversible is identified as Q_R, with the maximum work represented by W_a:

$$W_a = Q_R - \Delta H \tag{1.1}$$

Recognize that the fluid undergoes a reduction in enthalpy, such that $-\Delta H$ is, in fact, positive. By convention, work done by a fluid is positive whereas heat released by it is negative.

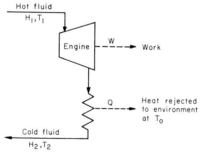

Fig. 1.1.   Only part of the energy in a hot fluid is convertible to work—the remainder must ultimately be rejected to the environment as heat.

The minimum amount of energy which must be given up as heat is related to the entropy change which the fluid undergoes:

$$dQ_R = T \, dS_R \tag{1.2}$$

This relationship is expressed in differential form because the absolute temperature T of the fluid changes continuously. In a purely reversible process, the entropy of the system including both the working fluid and the environment is constant. Therefore, their combined changes total zero:

$$\Delta S_R + \Delta S_0 = 0 \tag{1.3}$$

Because the environment is assumed to have a constant absolute temperature T_0, its increase in entropy varies directly with the amount of heat absorbed, Q_0:

$$\Delta S_0 = Q_0/T_0 = -Q_R/T_0 \tag{1.4}$$

Substituting (1.3) and (1.4) into (1.1) yields the available work as a function of the entropy change in the fluid:

$$W_a = T_0 \Delta S_R - \Delta H \tag{1.5}$$

The available work in a perfect gas as a function of its temperature T_1 elevated above the environment can be developed to illustrate the significance of temperature. If the gas gives up all its enthalpy, coming into equilibrium with the environment at T_0, then

$$\Delta H = C_p(T_0 - T_1) \tag{1.6}$$

where C_p is the heat capacity of the gas at constant (atmospheric) pressure.

The entropy lost by the gas in cooling to the temperature of the environment is the integral of Eq. (1.2):

$$\Delta S_R = \int \frac{dQ_R}{T} \tag{1.7}$$

Since

$$dQ_R = C_p dT \tag{1.8}$$

Eq. (1.3) is solved as

$$\Delta S_R = \int_{T_1}^{T_0} C_p \frac{dT}{T} = C_p \ln \frac{T_0}{T_1} \tag{1.9}$$

By combining (1.5), (1.6), and (1.9), it is possible to determine the theoretical maximum conversion of heat into work as the ratio of W_a to $-\Delta H$:

$$W_a/-\Delta H = 1 + [T_0/(T_1 - T_0)] \ln T_0/T_1 \tag{1.10}$$

The solution to Eq. (1.10) as a function of temperature difference of the hot fluid above an environment at 60°F (520°R) is given in Fig. 1.2.

Recognize that this curve represents the *maximum* amount of work attainable from an ideal gas at constant pressure, applied to a reversible heat engine. All real fluids and real processes will produce less work. It will be most useful in estimating the available work in the products of combustion which are used to drive a turbine, heat a boiler, etc. When a real working fluid such as steam or propane is encountered, more exact estimates of available work can be made using tabular data of its thermodynamic

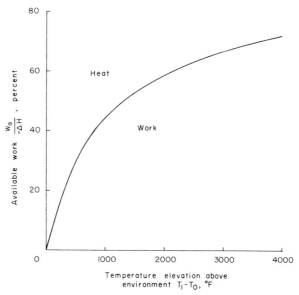

Fig. 1.2.   This curve describes the maximum work that can be recovered from an ideal gas in cooling to an environment at 60°F.

properties. Nonetheless, Fig. 1.2 clearly describes the value assignable to high-level (i.e., high-temperature) energy sources. Maximum plant efficiency can only be achieved when energy is allocated in such a way as to match its availability to the needs of the process.

2.   Entropy as a Measure of Irreversibility

Friction in the mechanical parts of the heat engine irreversibly converts shaft work into heat. Then the actual work delivered by the engine is less than the available work of the fluid:

$$W = Q - \Delta H \tag{1.11}$$

This is a restatement of Eq. (1.1), with actual work and heat substituted for available work and reversible heat.

If the initial and final states of the fluid are the same as when the heat engine operated reversibly, the entropy lost by the fluid is also the same— entropy is a property of state. Therefore, the increase in Q above Q_R indicates that the entropy of the environment has changed more than the entropy of the fluid, giving a net increase in system entropy S:

$$\Delta S = \Delta S_0 + \Delta S_R > 0 \tag{1.12}$$

and the work delivered is a function of the entropy gained by the environment:

$$W = -T_0 \Delta S_0 - \Delta H \tag{1.13}$$

Consequently the loss in available work is directly proportional to the net increase in the entropy of the system:

$$W_a - W = T_0 \Delta S \tag{1.14}$$

The effect of friction can then be quantified as an increase in system entropy. Conservation of available work therefore consists in minimizing the entropy increase in any process, hence minimizing its irreversibility.

3. Entropy and Disorder

Entropy is also considered a measure of disorder in the physical world. Physical systems have a natural tendency to equilibrate. Mountains fall, valleys fill, water flows downhill, vapors escape, materials corrode and decompose—all physical systems tend toward random distribution and disorder. These processes are all irreversible, increasing the entropy of the world. Only living systems are capable of proceeding from disorder to order. The photosynthesis of water and carbon dioxide into vegetable material, which is subsequently digested by animals into tissue, bone, and waste, are examples of living systems creating order out of randomness. Both energy and intelligence are needed to reduce entropy. A bottle broken by accident or mischief represents a natural irreversible process. Constructing or reconstructing a bottle requires a plan, skill, and energy as both work and heat.

Our reserves of fuels such as coal, oil, and wood were made by living systems (vegetable matter) drawing on solar energy. So ultimately all our energy came from and will come from the sun, and in turn be reradiated into space. The entropy of the world continues to increase as the energy from the sun is distributed into space.

Much of our fuel is expended in creating or restoring order—construction, agriculture, synthesis, refining, mining, fabrication, weaving—the examples are countless. These processes will use fuel most efficiently as they approach reversibility, and as their products are used reversibly. For example, aluminum is made from its ore in an electrolytic process. If scrap aluminum (such as cans) is simply discarded or buried, the electrical work put into forming the metal is lost. If the scrap is recycled, only a small fraction of the original energy must be used to return it to mint condition. The aluminum was given order in being formed from the ore; that order is entirely lost if the metal is mixed with other debris in a landfill. However, most of the order can be retained by returning the used metal for reconstitution. Recycling efforts such as this can save huge quantities of fuel, and must be pursued on a worldwide basis.

4. Identifying Irreversible Processes

Irreversible processes are characterized by the free fall, and the slip from order into disorder. Every expenditure of energy in which work is not recovered is irreversible. Flow always proceeds from a higher to a lower level. If work is recovered from that flow, then the flow can be reversed—at least in part—by applied work. But if no work is extracted by that flow, the process is completely irreversible.

A body falls reversibly in proportion to how much work is recovered from its loss in potential energy. If it falls freely, with its energy dissipated as heat, sound, fracturing, etc., the process is completely irreversible. If it lifts an equal weight an equal distance, the process is completely reversible—but we know that it cannot happen. Even in the absence of friction, there would be no driving force to produce motion. Hence, truly reversible processes must not only be ideal, they must also proceed at zero rate.

By this token, irreversibility increases with rate. For a given resistance, rate increases with driving force; consequently, irreversibility also varies directly with driving force. While Fig. 1.2 indicates that the percentage of available work in a perfect gas increased with driving force ($T_1 - T_0$), this curve describes a reversible process. Associated with any real process are irreversibilities, all of which increase with increased driving force. As a result, the work lost by increasing entropy increased with driving force, such that an *actual-work* curve will tend to depart more from the *available-work* curve as $T_1 - T_0$ increases.

An ordered state has bodies and fluids isolated and therefore unable to interact with each other. When released, elevated bodies tend to fall, vapors tend to escape, and fluids tend to blend—all without doing work. This tendency to seek a common level without doing work is characteristic of all systems lacking intelligence. The change from order to complete disorder (equilibrium) maximizes the entropy of the system. All processes which approach equilibrium without doing work are then irreversible. There are three general classes of irreversible operations found in the process industries:

(1) mixing of fluids having different compositions or temperatures,
(2) transfer of heat,
(3) flow of fluids through resistive devices.

Each of these operations is individually described below, with the increase in entropy and hence the loss in available work expressed mathematically.

B.   MIXING IS IRREVERSIBLE

Picture a balloon filled with helium, rising in the air. The balloon is punctured and the helium escapes. The irreversible nature of this process is revealed by the sound accompanying the release of gas. Yet there is a second irreversible process taking place which does not necessarily contribute to the sound—the mixing of helium with air. That mixing is irreversible should be obvious in that it produces no work, while its antithesis—separation—requires work.

Distillation is used extensively to separate and purify volatile liquids. In most distillation units, the raw feed and the refined products ultimately enter and leave at the same state. The refined products contain no more energy than the feed, yet energy is needed to make the separation. The intrinsic value of each product then resides in its composition, and energy is needed to obtain that value.

The antithesis of separation is mixing. If work is required to make a separation, then work is quantitatively lost whenever fluids having different compositions are combined. This loss in work may be determined from the entropy of mixing.

1.   The Entropy of a Mixture

Denbigh (1) gives the entropy of a mixture of ideal gases as

$$S = \frac{1}{F} \sum F_i (C_p \ln T - R \ln p_i) \tag{1.15}$$

where S is the entropy per mol of mixture, F the mols of mixture, F_i the mols of any component i in mixture, C_p the specific heat at constant pressure for component i, T the absolute temperature, R the gas constant, and p_i the partial pressure of component i. In an ideal-gas mixture, the mol fraction x_i of each component is the ratio of its partial pressure to total pressure p. Then we may substitute

$$p_i = x_i p \tag{1.16}$$

which gives

$$S = \frac{1}{F} \sum F_i (C_p \ln T - R \ln x_i - R \ln p) \tag{1.17}$$

Furthermore, the number of mols of each component F_i equals its mol fraction multiplied by the total mols of the mixture:

$$F_i = x_i F \tag{1.18}$$

When this expression is substituted into (1.17), we have

$$S = \sum x_i(C_p \ln T - R \ln x_i - R \ln p) \qquad (1.19)$$

The contribution of the composition itself to the entropy of the mixture is singled out to evaluate the effects of mixing under conditions of constant temperature and pressure:

$$S_x = -R\sum x_i \ln x_i \qquad (1.20)$$

Observe the double-valued function $x_i \ln x_i$, which is common to many thermodynamic and also statistical relationships [see Eq. (1.10)]. As x_i approaches zero and unity, the function approaches zero. It is plotted in Fig. 1.3, along with its sum for a binary mixture.

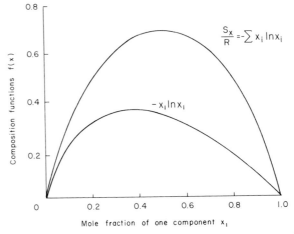

Fig. 1.3. The entropy of a binary mixture changes rapidly with composition at either extreme, reaching a maximum at equimolarity.

While the maximum entropy is reached in an equimolar mixture, the change in entropy with composition is most pronounced as purity is approached. Equation (1.20), evaluated for the binary system

$$S_x = -R[x \ln x + (1 - x)\ln(1 - x)] \qquad (1.21)$$

is differentiated to give

$$dS_x/dx = -R \ln[x/(1 - x)] \qquad (1.22)$$

The derivative is evaluated for conditions approaching absolute purity in Table 1.1. This relationship should indicate the work required to improve the purity of a product, and the ease with which it is contaminated.

Table 1.1

**The Change in Entropy of a Binary
Mixture as a Function of Purity**

x	S_x/R	$dS_x/R\ dx$
0.10	0.325	2.20
0.01	0.056	4.96
10^{-3}	7.91×10^{-3}	6.91
10^{-4}	1.02×10^{-3}	9.21

Multicomponent mixtures cannot be described graphically as easily as a binary system because of their many degrees of freedom. Nonetheless, the effects of additional components can be appreciated by observing the maximum value of the function $-\sum x_i \ln x_i$, which occurs in equimolar mixtures. Table 1.2 lists these values for mixtures up to 10 components.

Table 1.2

**The Maximum Values for the Function
$-\sum x_i \ln x_i$ for (Equimolar)
Multicomponent Mixtures**

n	$-\sum x_i \ln x_i$	n	$-\sum x_i \ln x_i$
2	0.693	5	1.609
3	1.099	7	1.946
4	1.386	10	2.302

2. Mixing for Quality Control

Controlling the composition of products leaving a separation process such as distillation, by adjusting its material and/or energy balance, is difficult. The distribution of internal capacities causes severe delays in response, such that much effort has been expended in improving control (2). By contrast, quality control by mixing is responsive and predictable. Consequently, there is a temptation to mix overpure and underpure products in order to exactly meet a given specification.

Mixing may be practiced in a separations unit in a number of ways. Raw or partially refined material can be allowed to bypass the unit to adjust the composition of an overly purified product. A second possibility is the mixing of products from two parallel units to average their purity. Or the product from a single unit may be mixed with material of a different composition that was produced earlier. Also, products that have failed to meet specifications are customarily mixed with raw feed for reprocessing. Occasionally, recycle streams containing partially purified products are mixed continuously with feedstocks. And the practice of introducing feedstock whose composition fails to match that of the mixture existing at the point of introduction is very common.

When streams are mixed, the composition of the mixture is the weighted average (represented by the bar) of the compositions of the streams in mass or molar units:

$$\bar{x}_i = \sum F_j x_{ij} / \sum F_j \tag{1.23}$$

where $\bar{x}_i$ is the fraction of any component i in the mixture and x_{ij} is its fraction in a stream j which is flowing at rate F_j. However, the entropy of the mixture must always be greater than the weighted average of the entropy of the individual streams:

$$\bar{S} > \sum F_j S_j / \sum F_j \tag{1.24}$$

The increase in the entropy of the system brought about by mixing represents a loss in available work, i.e., a loss in the work used to attain the original composition,

$$\Delta S = \bar{S} - \sum F_j S_j / \sum F_j \tag{1.25}$$

which can be restated using compositions,

$$\Delta S = -R\left[\sum \bar{x}_i \ln \bar{x}_i - \sum (F_j \sum x_{ij} \ln x_{ij}) / \sum F_j\right] \tag{1.26}$$

Because of the nonlinear nature of the curves in Fig. 1.3, ΔS increases with the difference in the compositions between the mixture and the feedstreams. Table 1.3 cites ΔS for three different combinations of binary streams which can be blended to form the same mixture.

If mixing is necessary, the above example indicates that the least loss in work will be incurred when the compositions of the original streams are as

Table 1.3

The Entropy Increase Sustained by Mixing to Reach $\bar{x} = 0.02$, Using Different Feedstocks

x_1	$\dfrac{F_1}{F_1 + F_2}$	x_2	$\dfrac{F_2}{F_1 + F_2}$	$\Delta S/R$	$\Delta S/\bar{S}$ (%)
0.01	0.50	0.03	0.50	0.0026	3.3
0.00	0.50	0.04	0.50	0.0140	17.9
0.01	0.98	0.50	0.02	0.0290	37.1

close together as possible. In the case of a product from a separations unit whose composition is cycling, ΔS increases with the amplitude of that cycle, as overpure and underpure products are combined in the receiving tank. Mixing the products from parallel units would seem to incur relatively little penalty, whereas bypassing the separations unit with even a miniscule amount of raw feed increases entropy considerably. Recycling product back to the feed obviously wastes much of the work used to effect the separation. Bojnowski et al. (3) reduced the energy consumed in a distillation system by 67% after eliminating four such recycle streams. But the maximum increase in entropy possible by mixing occurs when pure streams are combined, in which case $\Delta S = S_x$. This practice wastes *all* of the energy put into purifying the streams.

3.   Mixing for Temperature Control

Mixing is also commonly used to achieve temperature control, in a manner described in Fig. 1.4. Not only is this practiced in industrial plants, it is also familiar at home, where hot and cold water are mixed for washing. For a given flow of heat delivered to the users, the minimum amount of fuel will be used when the bypass is closed. This relationship can be demonstrated from both an entropy and a heat-transfer standpoint.

For a pure ideal gas at atmospheric pressure, Eq. (1.15) reduces to

$$S = C_p \ln T \tag{1.27}$$

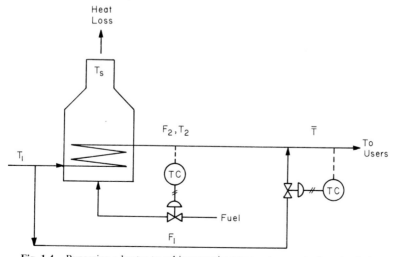

Fig. 1.4.   Bypassing a heater to achieve precise temperature control wastes fuel.

If streams of the same ideal gas at different temperatures are mixed, the temperature of the mixture becomes

$$\bar{T} = \sum F_j T_j / \sum F_j \tag{1.28}$$

where F_j is the flow rate of each individual stream whose temperature is T_j.

As the temperatures of the two streams differ, the entropy of the mixture will exceed the sum of the entropies of the original streams by the entropy of mixing:

$$\Delta S = \bar{S} - \frac{\sum F_j S_j}{\sum F_j} = C_p \left[\ln \bar{T} - \frac{\sum (F_j \ln T_j)}{\sum F_j} \right] \tag{1.29}$$

Following the example described in Table 1.3, Table 1.4 gives the entropy increase caused by mixing various combinations of hot and cold streams to arrive at a uniform temperature. Refer to Fig. 1.4 to identify the streams. The last column shows the increase in entropy due to mixing as a percentage of the increase brought about by heating the fluid from its initial to its final delivered temperature. This additional entropy represents available work that was lost—additional fuel needed to fire the heater must be supplied in *at least* the percentage appearing in the last column. As can be seen, the increase in entropy is almost directly proportional to the bypass flow rate.

Table 1.4

The Entropy Increase Sustained by Mixing to Reach $\bar{T} = 400°$F (860°R) with Variable Bypassing

T_1 (°F)	T_2 (°F)	$\dfrac{F_1}{F_1 + F_2}$	$\Delta S / C_p$	$\Delta S / (\bar{S} - S_1)$ (%)
100	450	0.143	0.0128	2.99
	500	0.250	0.0247	5.77
	600	0.400	0.0461	10.75

Let the heat transferred to the product be related to the flue gas temperature T_s by a simple convection equation, where the heat-transfer coefficient varies with the 0.8 power of flow:

$$Q = kF_2^{0.8}(T_s - T_2) \tag{1.30}$$

When F_1 is zero, T_2 and therefore T_s are at their minimum (subscript m), and F_2 is at its maximum value of $F_1 + F_2$:

$$Q = k(F_1 + F_2)^{0.8}(T_{sm} - T_{2m})$$

Since the heat transferred to the product is the same in any case, the last two expressions are equal, such that

$$T_s = T_2 + (T_{sm} - T_{2m})[(F_1 + F_2)/F_2]^{0.8} \qquad (1.31)$$

Additional heat lost from the stack due to bypassing is proportional to $T_s - T_{sm}$. Table 1.5 relates this additional heat loss to the heat transferred to the product (as the difference between 3000°F flame temperature and T_s), assuming 600°F for T_{sm}.

While the heat-transfer equations included some assumptions which may be inaccurate, the last columns in Tables 1.4 and 1.5 are in general agreement. Bypassing the heater raises T_2 and reduces F_2, *both* effects promoting stack losses.

Table 1.5

**Incremental Heat Lost in the Flue Gas
Due to Mixing**

$\dfrac{F_1}{F_1 + F_2}$	T_2 (°F)	T_s (°F)	$\dfrac{T_s - T_{sm}}{3000 - T_s}$ (%)
0.0	400	600	0.0
0.143	450	676	3.2
0.250	500	752	6.8
0.400	600	901	14.3

C. HEAT TRANSFER IS IRREVERSIBLE

The working fluid in a heat engine must receive heat to raise it to its maximum temperature, and reject heat allowing it to cool to its minimum temperature. The difference between the heat received and the heat rejected by the fluid is the work produced by the engine. Because a certain temperature drop must be established across a heat-transfer surface proportional to heat flow, the working fluid can never be as hot as the heat source, nor as cool as the environment. Consequently, the work obtainable from a hot fluid will always be less than its available work in proportion to those temperature differences. This loss in available work is due to the irreversibility associated with heat transfer, and is directly calculable as an increase in entropy.

1. The Entropy of Heat Transfer

In the reversible heat–work cycle, the entropy gained by the environment in removing heat from the working fluid was exactly equal to the entropy lost by the fluid. But this is only possible when the temperature difference between the working fluid and the environment is zero. This relationship can be demonstrated by considering how saturated steam might be condensed by boiling exactly the same amount of water at the same pressure. The steam would give up the same amount of enthalpy and entropy as the water would absorb—but only if their pressures (and temperatures) were identical.

But heat transfer between two fluids requires a finite temperature difference. As a result, the cooler fluid in absorbing the same amount of heat given up by the hot fluid will have its entropy increased more than the entropy of the hot fluid is reduced. Consider a fluid at T_1 transferring a differential amount of heat dQ to the environment at T_0. The loss in entropy of the hot fluid is

$$dS_1 = dQ/T_1 \tag{1.32}$$

while the gain in entropy of the environment is

$$dS_0 = -dQ/T_0 \tag{1.33}$$

The net change in entropy for the system is

$$dS = dS_0 + dS_1 = -dQ(1/T_0 - 1/T_1) \tag{1.34}$$

Only if T_0 and T_1 are equal is $dS = 0$; therefore, the irreversibility of the heat-transfer process is related to the temperature difference between the fluids.

In the special case of isothermal heat transfer between two fluids, both of which are changing state, the integral of (1.34) approaches

$$\Delta S = -Q(1/T_0 - 1/T_1) \tag{1.35}$$

Consider the example of steam at 10 psia condensing against water boiling at 8 psia. The heat given up in condensing at 10 psia is 982.1 Btu/lb, whereas the heat absorbed in vaporizing water at 8 psia is 988.5 Btu/lb. Neglecting heat losses, 0.9935 lb of water vaporizes per pound of steam condensed. The entropy of condensation at 10 psia is -1.5030 Btu/lb-°F, while that of vaporization at 8 psia is $+1.5384$ Btu/lb-°F. The entropy increase in the system due to heat transfer is

$$\Delta S = 0.9935(1.5384) - 1.503 = 0.0254 \quad \text{Btu/lb-°F}$$

Using Eq. (1.35), the entropy increase can be estimated from the condensing temperature of 193.21°F at 10 psia and the boiling point of 182.86°F at 8 psia:

$$\Delta S = 982.1 \left(\frac{1}{460 + 182.86} - \frac{1}{460 + 193.21} \right) = 0.0242 \quad \text{Btu/lb-°F}$$

The fact that the increase in entropy for the real fluid (steam) is greater than that found for an ideal fluid under the same conditions using Eq. (1.35) can be ascribed to the nonideality of the real fluid. Every departure from ideality will result in an increase in entropy and therefore a reduction in available work from what theoretical considerations have predicted.

Nonisothermal heat transfer results in still greater increases in entropy, since entropy depends on temperature. Consider condensing steam at 10 psia against cooling water. The water temperature will rise such that ΔS_0 must be calculated by integrating Eq. (1.33) between the limits of its inlet T_{01} and outlet T_{02} temperatures:

$$\Delta S_0 = -\int_{T_{01}}^{T_{02}} \frac{dQ}{T_0} = FC_p \ln \frac{T_{02}}{T_{01}} \tag{1.36}$$

where F is the mass of coolant of specific heat C_p undergoing the change in temperature. A heat balance on the system gives

$$-Q = FC_p(T_{02} - T_{01}) \tag{1.37}$$

which allows elimination of FC_p from (1.36):

$$\Delta S_0 = -Q[\ln(T_{02}/T_{01})/(T_{02} - T_{01})] \tag{1.38}$$

Then the net entropy increase for the system becomes

$$\Delta S = -Q\left[\frac{\ln(T_{02}/T_{01})}{T_{02} - T_{01}} - \frac{1}{T_1}\right] \tag{1.39}$$

As T_{02} approaches T_{01}, the left-hand term within the brackets approaches $1/T_{01}$. Table 1.6 indicates the departure from $1/T_{01}$ as T_{02} increases above T_{01}. Recognize that Table 1.6 shows only part of the penalty due to non-isothermal heat transfer. As the coolant temperature rises, the condensing

Table 1.6

Entropy Increase Due to Nonisothermal Heat Transfer

T_{01} (°R)	$T_{02} - T_{01}$ (°F)	$\frac{\ln(T_{02}/T_{01})}{T_{02} - T_{01}}(\times 10^{-3})$	$\frac{1}{T_{01}}(\times 10^{-3})$	$\frac{1}{T_{02}}(\times 10^{-3})$
520	1	1.921	1.923	1.919
	2	1.919		1.916
	5	1.914		1.905
	10	1.905		1.887
	20	1.887		1.852
	50	1.836		1.754

temperature T_1 for the working fluid also rises, further increasing the entropy of the system.

As the heat load on the condenser is increased, ΔS will rise with Q, with $T_1 - T_0$, and with $T_{02} - T_{01}$. Consequently, the irreversibility of any uncontrolled heat-transfer process increases sharply with heat load—reversibility is approached at no load. However, the addition of controls to regulate temperatures or pressures tends to restrict heat transfer at lower loads by throttling, flooding, etc., thereby artificially maintaining a high degree of irreversibility.

Figure 1.5 shows examples of coolant throttling and condensate flooding used to control the pressure within a condenser. As long as the heat load is below the maximum that the condenser can carry at the stipulated pressure, heat transfer must be impeded by control action. However, entropy conservation requires that control not be applied—then the process will be able to reject heat as reversibly as possible, thereby maximizing its thermodynamic efficiency. This concept is explored in more detail in later chapters on refrigeration and distillation.

While the discussion to this point has centered on heat transfer at the cold end of the cycle, irreversibility at the hot end is even worse. Heat transfer between combustion products and the working fluid (e.g., steam) is conducted across such a large temperature gradient that substantial work is lost. Consider generating superheated steam at 1000°F by burning oil at a flame temperature of 3000°F. From Fig. 1.2, the available work in the products of combustion at 3000°F is about 66% while that in the steam at 1000°F is about 44%. (The available work in steam differs from that in a perfect gas, although with a high degree of superheat the difference is small.)

Or consider the combustion products being sent to a gas turbine whose high-temperature limit is 2000°F. Excess air would have to be added to control the turbine inlet temperature at 2000°F, thereby reducing the

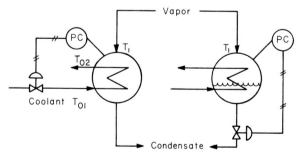

Fig. 1.5. Condenser pressure can be controlled by throttling the coolant or flooding the heat-transfer surface—both practices raise the entropy of the system.

available work of the fluid to 59%. Although control in this case is achieved by blending rather than heat transfer, the effect is the same. Conditions at the hot end of the process will always contribute most to entropy increase because of temperature limitations in heat-transfer tubing and rotating machinery.

2. Use and Reuse of Energy

Perhaps the greatest opportunity for energy conservation in industry lies in the multiple use of energy. The fluids having the highest temperatures and pressures in a plant also have the highest percentage of available work. These fluids should then be used to drive expansion engines (turbines) to extract at least part of their available work as mechanical energy for pumps and compressors or to generate electricity. Rather than allow these engines to exhaust to a condenser, however, the heat remaining in the fluid should be used for process heat. In a perfectly balanced plant, no electricity would be purchased, and no steam would have to be condensed against cooling water or air.

A brief analysis of this last concept is in order. If the plant buys electricity, that electricity is probably generated by a condensing turbine. Because the power plant only produces work and discards its heat, its overall efficiency cannot be as great as a plant whose work and heat are *both* used. If the process plant uses more work than heat and does not buy electricity, however, it will have to either condense or vent its low-pressure steam. If its demand for heat exceeds the demand for work, then the additional heat should be provided by generating and selling electricity to users incapable of generating their own. The problem of balancing work and heat is developed in detail in Chapter 3. At this point, it is sufficient to conclude that work should be extracted from high-level energy sources wherever possible.

From this standpoint, combustion of fuel that is used only for heat is inconsistent with conservation of work. Furthermore, the conversion of work into heat, as in electrical resistance heating, is counterproductive, and to be avoided whenever possible.

The conversion of heat into mechanical work and thence into electrical energy is the most visible example of attaching value to heat, but is far from being alone. Scattered throughout industry are a variety of operations which use energy, although without it appearing in their products. This concept was touched upon earlier in the discussion on the entropy of mixing. The products from any separations unit generally have the same energy content as the feedstock—yet, because they are purer, their entropy is lower. Their entropy is reduced by increasing the entropy of the heat-transfer media used to effect the separation.

Because the energy used and rejected by the process is essentially the same quantity, it is capable of reuse, depending on its temperature. This technique has been practiced for years in concentrating solutions through multiple-effect evaporation. Vapor generated by condensing steam in the first effect is employed as the heating medium in the second effect, etc., as shown in Fig. 1.6. The most common installations have three effects: steam at perhaps 20 psia is supplied to the first, producing vapors at 10, 5, and 2 psia, in subsequent effects. Absolute pressure in the last effect is maintained with a water-cooled condenser; a vacuum pump exhausts noncondensible gases.

Usually the solution is aqueous. Then the mass of vapor removed per unit steam is somewhat less than the number of effects, due to heat losses and latent heat of vaporization increasing as pressure decreases. For a given steam supply pressure and coolant temperature, the number of effects and hence the conversion of steam to vapor is determined by the temperature drop across each heat-transfer surface. If temperature drops are reduced by expanding heat-transfer surfaces, more effects can be employed and consequently greater economy can be achieved. But the temperature differences are due not only to heat flow but also to boiling-point elevation due to solids concentration, and static and velocity heads. These real considerations determine the optimum heat-transfer surface and hence the optimum number of effects.

If steam is supplied at a higher pressure, however, the number of effects and hence the economy of the system could increase substantially. This procedure cannot be extended without limit, due to the temperature sensitivity of many products. Nonetheless, in principle, steam at a very high pressure could provide many more effects of evaporation than steam at low pressure. This demonstrates that the generation of low-pressure steam by the combustion of fuel may result in much more fuel consumption than if high-pressure steam were generated.

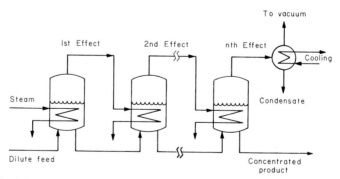

Fig. 1.6.   A multiple-effect evaporator uses the same energy several times over.

If the average temperature drop across each effect were about 32°F, 20-psia steam could supply a triple-effect evaporator as described. Neglecting heat losses, each pound of steam would drive off 2.88 lb of vapor. If the unit were supplied with steam at 140 psia, seven effects could be used. Neglecting heat losses, each pound of steam could then evaporate 6.37 lb of water. While heat losses and other considerations will reduce its attractiveness, the value of the higher-pressure steam is manifest in this example.

Applying the multiple-effect principle to other processes is not so easy. Parallel distillation columns may be operated at different pressures so that the heat rejected by one may reboil the other. Cascading of energy may also be applied in either a forward or backward path in the serial distillation of a multicomponent mixture. These examples are discussed in detail in the appropriate chapters. Here the point is made that energy can be used and reused in many common processes, given sufficient foresight, the availability of energy at high levels, and controls which coordinate such reuse. In every case, heat-transfer must be conducted at maximum reversibility, i.e., at minimum temperature drop. Again, controls should not interfere with this objective.

3.   An Energy Accounting System

If a plant is to be managed to conserve energy in whatever form it appears, a graduated accounting system is mandatory. Steam may be worth more than its enthalpy indicates, and the heat of combustion of a fuel does not accurately describe its value. The example cited above bears this out: 140-psia steam was capable of evaporating over twice as much water as 20-psia steam, although its enthalpy is but 3% higher. Similarly, a low-Btu fuel cannot produce the flame temperature of a higher-valued fuel, limiting the work which may be extracted from it.

A common denominator is needed to equate all forms of energy—*available work* is suggested as the accountable property. The available work in a fuel may be evaluated from its flame temperature when burned in air at 60°F; the absolute flame temperature would then be inserted as T_1 in Eq. (1.10) with 520°R (60°F) used as T_0. The relationship between flame temperature and fuel heating value is explored in Chapter 2. Reference (4) gives the maximum observed flame temperature of methane in air at 3416°F. Using Eq. (1.10) with $T_0 = 520°R$, the available work in its products of combustion is 68.9% of its higher heating value of 1013 Btu/ft³, or 697 Btu/ft³.

Steam can be supplied at a variety of pressures and temperatures. The available work for each condition can be determined by inserting the appropriate data from steam tables into Eq. (1.5). The reference temperature T_0 would typically be 520°R, and ΔS_R would be the entropy difference

Table 1.7

The Available Work in Steam at Various Conditions

Pressure (psia)	Temp. (°F)	Enthalpy (Btu/lb)	Entropy (Btu/lb-°F)	W_a (Btu/lb)	$-W_a/\Delta H$ (%)
2500	1000	1458	1.527	665	46.5
1200	1000	1449	1.630	602	42.4
600	700	1352	1.584	529	40.0
400	445[a]	1205	1.485	433	36.8
250	401[a]	1201	1.526	408	34.8
140	353[a]	1193	1.575	375	32.2
100	328[a]	1187	1.603	354	30.5
50	281[a]	1174	1.659	312	27.2
20	228[a]	1156	1.732	256	22.7

[a] Saturated.

between its initial conditions and water at the reference of 60°F (0.056 Btu/lb-°F). Similarly, ΔH would be the enthalpy change from initial conditions to water at 60°F (28 Btu/lb). Table 1.7 compiles the available work for several typical steam conditions. Observe that the percent availability of *saturated* steam exceeds that of an ideal gas at the same temperature, because it can release energy isothermally. Figure 1.7 relates available work to steam pressure for conditions of saturation and 1000°F, on a logarithmic pressure scale.

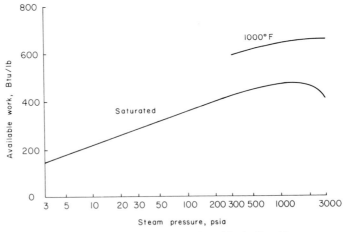

Fig. 1.7. The available work in steam increases logarithmically with pressure up to 1500 psia, and also increases with temperature.

It is meaningful to compare steam at various pressures and temperatures on the basis of available work, and one fuel against another. Yet steam cannot be compared to fuel or electricity on the basis of available work alone. There are several reasons for this. Conversion of fuel to steam results in stack losses and a reduction in available work as well. Furthermore, conversion of steam to electricity encounters irreversibilities in heat transfer and nonisentropic expansion. Additionally, power distribution losses must be taken into account.

The "heat rate" of central-station power plants fired by fossil fuels ranges from 9000 to 11,000 Btu of fuel consumed per kilowatt hour of electricity generated. Using 10,000 Btu/kWh as a mean, the cost of electrical energy ought to be 10,000/3413 or 2.9 times the cost of the energy in fossil fuels. Electrical rates generally bear this out, being three to five times as costly as fuels, depending on the particular rate structure.

Because cost information is readily available for purchased electricity, there is generally no misunderstanding regarding its value, either purchased or generated. Confusion arises about the value applied to steam and to fuels relative to their quality. Individual fuels can be discriminated among one another on the basis of available work, and steam sources can be treated similarly. Then to convert from the work in electricity to that in steam or fuel requires a conversion factor. If electricity is valued at 2.9 times the energy content of fuel, then it could be valued at 2.0 times the available work in methane gas, based on the percentage available work in the gas.

The conversion factor from fuel to steam is determined only in part by the thermal efficiency of the boiler. Consider a 400-psia saturated-steam boiler with a thermal efficiency of 75%. A single Btu of fuel would produce 0.75 Btu of steam enthalpy (above water at 60°F) but only 0.276 Btu of available work. Compared to the 68.9% available work in the fuel, the work efficiency of the boiler would be 0.276/0.689 or 40%. Thus low-pressure boilers would tend to have much lower work efficiencies than high-pressure boilers.

D.   THROTTLING IS IRREVERSIBLE

In order to control a plant it is necessary to modulate its flowing streams, either stepwise with on–off devices, or continuously by throttling devices. Because on–off control is rarely satisfactory, being incapable of reaching a steady state, most regulation is exercised through throttling.

Throttling is defined as a reduction in pressure accomplished without the removal of energy, either as heat or work. It is important that throttling be achieved with a minimum of energy loss—recognize that energy lost here will have a destructive influence on the internals of the throttling mechanism.

As a consequence, users of throttling devices may have considered them to be loss-free, hence bearing little importance to any energy-conservation effort. This is, of course, false. While they may not waste energy directly, they lose available work, with the result that the energy passed is less useful. Ultimately, the result is wasted energy.

1.   Isenthalpic versus Isentropic Expansion

The pressure in a stream may be reduced with or without the extraction of work. If no work is extracted, and no heat is lost, the pressure reduction is isenthalpic, in that the enthalpy of the fluid is unchanged. If the enthalpy of the fluid is reduced by extracting work reversibly, the pressure reduction is isentropic. In this case, the maximum amount of work is extracted, in perfect contrast to the isenthalpic case. In any real engine, friction and fluid nonidealities will limit the reversibility of the process, so that less than the maximum work may be extracted. The efficiency of converting available work to actual work is a measure of the reversibility of the process. This concept was demonstrated by Eq. (1.14).

The degree of reversibility in an expansion process can be seen by referring to a temperature–entropy diagram such as Fig. 1.8. Steam expanding reversibly through an engine will follow a vertical downward path of constant entropy. Observe that superheat is lost during the expansion, which can continue into the mixed-phase region. Therefore, a turbine which is expected to extract a maximum amount of energy from the steam must be capable of discharging a certain percentage of liquid.

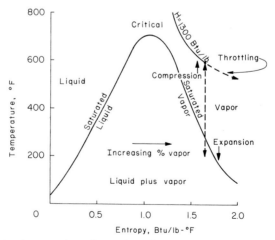

Fig. 1.8. Isentropic expansion of superheated steam can cause condensation, whereas isenthalpic expansion (throttling) increases superheat.

Any irreversibility will cause the expansion path to slope to the right. When the discharge pressure corresponding to the temperature of the heat sink is reached, less enthalpy will have been lost to the engine than in the reversible expansion. The loss in enthalpy is still equal to the amount of work extracted, to satisfy the energy balance across the engine, but less work is extracted. A real turbine will have an efficiency factor of 70–80%, meaning that it is capable of extracting 70–80% of the enthalpy which the steam would lose in isentropic expansion.

Expansion through an orifice or throttling device follows a contour of constant enthalpy. For a given pressure drop, temperature falls only slightly compared to isentropic expansion. Because its temperature falls much less than the pressure, steam expanding isenthalpically becomes more super-heated, in contrast to isentropic expansion, where superheat decreases. Superheated steam is a poor heat-transfer medium, having a relatively low film coefficient, and cannot transfer heat isothermally. Yet saturated and wet steam are excellent for heat transfer, having a high coefficient along with isothermal heat transfer. Consequently, turbine-exhaust steam is excellent for process heat. On the other hand, steam that has been let down to a reduced pressure through a regulator or control valve is superheated and therefore not as valuable for heating.

Isentropic compression follows the same vertical line as expansion be-cause one of these processes is the reverse of the other. (There is no reverse of isenthalpic expansion because it is an irreversible process.) Just as expansion through an engine reduces superheat, compression increases it. Furthermore, some compressors are designed to accept vapors containing a few percent liquid, just as turbines can exhaust vapor with a few percent liquid. Com-pressors are typically 70–75% efficient, such that their path deviates from the vertical in the T–S diagram slightly to the right (increasing entropy). As with expansion, all the work applied to the compressor appears in the enthalpy of the fluid, although not all as available work of the compressed fluid, due to its increase in entropy.

When the working fluid for an engine is *saturated* steam, substantial amount of moisture will condense in the turbine unless its efficiency is low. If no vapor were to condense, the expansion path would have to follow the saturated-vapor contour, at considerable loss in work. Most turbines designed for saturated steam exhaust some moisture, such that their expan-sion path lies between the saturated-vapor curve and the vertical.

The maximum efficiency of a saturated-vapor Rankine cycle will be realized when the turbine exhaust is also saturated. This requires that the saturated-vapor curve slope negatively enough to match the turbine efficiency in the operating region, as shown in Fig. 1.9. Some organic fluids have a positive-sloping saturated-vapor curves, also appearing in Fig. 1.9. Their

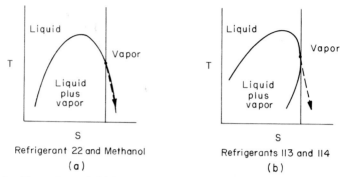

Fig. 1.9. The optimum fluid for a Rankine cycle will have a saturated-vapor curve that matches the expansion curve for the engine (a). Most organic fluids exhibit a positive slope (b) and will be exhausted in a superheated state.

vapors will leave an engine superheated and hence will not condense efficiently. Refrigerants 113 and 114 have this positive slope (5). Others, such as Refrigerant 11 and propane have a nearly vertical slope, while still others such as Refrigerant 22 and methanol slope slightly negative.

On the *liquid* side of the *T–S* diagram, pumping and flashing can be illustrated as in Fig. 1.10. Raising liquid pressure isentropically by a pump causes the liquid to be subcooled at its higher pressure. Expansion of saturated liquid through an engine releases a certain amount of vapor. Because of the large difference in volume between liquid and vapor, flashing of vapor in an engine increases local velocities to the point where two-phase flow can

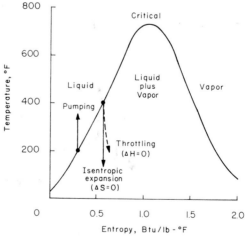

Fig. 1.10 Isenthalpic expansion (throttling) of saturated water flashes slightly more vapor than isentropic expansion, indicating a small loss in available work.

cause damage. Consequently, the recovery of power by depressuring a liquid cannot be carried out under conditions where much vapor will be formed.

If a saturated liquid is passed through a valve, more flashing to vapor will occur than in isentropic expansion as shown in Fig. 1.10, but a valve can accommodate vapor more easily than an engine. Consequently, refrigeration systems use a valve to expand the condensed refrigerant. If an expansion turbine could be used, some of the power of compression could be recovered and less of the refrigerant would flash to vapor. Thus the efficiency of the cycle could be improved considerably—unfortunately expanders designed for saturated liquids are not available.

Loss in available work in liquids can be estimated from the integral of the pressure–volume product. Because they are scarcely compressible, that product is simply

$$\Delta W_a = v \, \Delta p \tag{1.40}$$

When volume is expressed in cubic feet and pressure in pounds per square foot, their product is foot-pounds, which, multiplied by 1.286×10^{-3}, gives Btu. For a *flowing* system, the volumetric flow rate may be used for v, in which case ΔW_a represents a loss in available power.

The volume of a *compressible* fluid changes with pressure, however, so that in the isothermal expansion of an ideal gas, $\Delta(pv) = 0$. Then the loss in work can be determined on the basis on an entropy increase, per Eq. (1.14):

$$W_a - W = T_0 \, \Delta S \tag{1.14}$$

When a gas is expanded isenthalpically, its temperature may change according to the Joule–Thompson coefficient μ:

$$\mu \equiv \frac{\partial T}{\partial p}\bigg|_H \tag{1.41}$$

However, in an ideal gas, $\mu = 0$, in which case temperature is constant. Then Eq. (1.15) may be differentiated for one mole of a pure gas to give

$$\frac{\partial S}{\partial p}\bigg|_T = -\frac{R}{p} \tag{1.42}$$

For isothermal expansion of an ideal gas, then

$$dS = -R\frac{dp}{p} \tag{1.43}$$

and

$$\Delta S = -R \ln(p_2/p_1) \tag{1.44}$$

where p_1 and p_2 are inlet and outlet pressures, respectively.

The isenthalpic expansion of steam at 1200 psia and 1000°F to 1000 psia can be used to illustrate this relationship. If it were an ideal gas,

$$\Delta S = -\frac{1.987}{18}\frac{\text{Btu/lb mol-}°R}{\text{lb/lb mol}}\ln\left(\frac{1000}{1200}\right) = 0.0201 \quad \text{Btu/lb-}°R$$

and

$$W_a - W = (520°R)\Delta S = 10.47 \quad \text{Btu/lb}$$

The steam tables indicate, however, that expansion is *not* isothermal, such that the final temperature reached is 989°F. Entropy data for those conditions indicate an increase of 0.020 Btu/lb-°R—essentially the same result as obtained for an ideal gas.

2. Selecting Control Valves

While many procedures have been written describing methods for sizing control valves, most begin with given inlet and outlet pressures. Only in rare instances are these pressures actually constant, and still less often are they selected for optimum performance of the process.

The control valve must work in concert with the piping, vessels, and prime mover in each stream. In general, the vessels are specified first, then the piping designed for an optimum combination of capital and operating costs. Next, the pump or compressor is selected *assuming* that the control valve will add a certain increment to the system pressure loss. The value of that increment is usually assigned on an arbitrary basis, from 25 to 33% being common. Once the prime mover is specified, the power lost through the control valve is already determined. Recognize that, at any given flow, the mover will develop a certain head, part of which is absorbed by the fixed resistances of the vessels and piping, the balance being taken across the valve. Changing the valve without changing the prime mover will then not affect the work lost by the valve.

The control valve needs to be sized *before* the prime mover, rather than later. And its allowable pressure drop needs to be determined on the basis of control rangeability, rather than being assigned arbitrarily. If too little drop is available across the valve, its effect on the process will vary appreciably with flow, and control can be inconsistent. Exactly where this point occurs depends on the rangeability required by the process and the installed characteristic of the control valve.

The flow of liquid passed by a control valve is related to pressure drop by the formula

$$F = aC_v\sqrt{\Delta p/\rho} \tag{1.45}$$

where F is the flow (gpm), a the fractional valve opening, C_v the flow coefficient, Δp the pressure drop (psi), and ρ the fluid density (g/ml). (A comparable formula exists for gases, taking into account their compressibility.) If Δp were constant, flow would be directly proportional to valve opening. However, Δp tends to decrease with flow squared, due to the resistance in vessels, piping, and pump internals. Then the pressure drop available to the control valve decreases from a maximum Δp_M at zero flow:

$$\Delta p = \Delta p_M - \rho(F/C_r)^2 \tag{1.46}$$

where C_r is the lumped flow coefficient of the fixed resistances. Equation (1.46) can also be written for maximum flow F_M corresponding to a wide-open valve, in which case Δp would be the minimum Δp_m. Dividing (1.46) evaluated at any flow F by itself evaluated at F_M eliminates ρ and C_r:

$$(F/F_M)^2 = (\Delta p_M - \Delta p)/(\Delta p_M - \Delta p_m) \tag{1.47}$$

Similarly, (1.45) can be written for F_M, in which case $a = 1$ and $\Delta p = \Delta p_m$. Dividing (1.45) by itself evaluated at F_M eliminates ρ and C_v:

$$(F/F_M)^2 = a^2 \, \Delta p/\Delta p_m \tag{1.48}$$

When Eqs. (1.47) and (1.48) are combined by eliminating Δp, the fractional flow F/F_M can be related to fractional valve opening a and the ratio of minimum to maximum pressure drop:

$$\frac{F}{F_M} = \frac{1}{\sqrt{1 + (\Delta p_m/\Delta p_M)(1/a^2 - 1)}} \tag{1.49}$$

Curves of fractional flow versus fractional valve opening are shown in Fig. 1.11.

If the valve has a linear characteristic, i.e., fractional valve opening a is equal to fractional stem position m, then Fig. 1.11 is illustrative of its *installed* characteristic. If a linear installed characteristic is desirable using a linear valve, then the valve would have to absorb most of the system pressure drop. From an energy conservation standpoint, this type of operation would be wasteful for most streams normally encountered in a plant. The most appropriate course is then to choose another valve characteristic which, when installed, will result in a more linear relationship between stem position and flow.

The valve chosen to perform this function has an "equal-percentage" characteristic. It is logarithmic in nature, having the following relationship between valve opening and stem position:

$$a = r^{(m-1)} \tag{1.50}$$

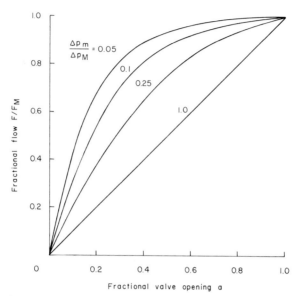

Fig. 1.11. The characteristic of a linear valve is distorted in inverse proportion to its fraction of total system head loss.

where r is the valve's stated rangeability. Manufacturers list rangeability ratings for their valves, with figures of 33 and 50 being common, although 500 and even 1000 have been cited. A plot of flow versus stem position for the parameters of Fig. 1.11 can be obtained by solving (1.50) for selected values of m, then solving (1.49) for values of a thus found. The results are plotted in Fig. 1.12 for a valve with a rangeability of 50.

Figure 1.12 indicates that an equal-percentage valve is capable of providing an installed characteristic that is nearly linear, when comparatively little pressure drop is allocated to it at maximum flow. However, the rangeability is reduced in proportion to the pressure-drop allocation. At full flow, the drop is only Δp_{m}, while near zero flow it increases to Δp_{M}. The installed rangeability of the valve is then

$$r_I = r\sqrt{\Delta p_{\mathrm{m}}/\Delta p_{\mathrm{M}}} \qquad (1.51)$$

For the valve plotted in Fig. 1.12, the rangeability is reduced from 50 to 11.2 when only 5% of the system drop is allocated at full flow. In most cases, this provides adequate control, however. Then a power savings of 25% can be realized by selecting the valve and prime mover for 5% allocated pressure drop versus a nominal 30%. Whether it is worth attempting to lower the allocated pressure drop below 5% is questionable—that may be the point of diminishing returns.

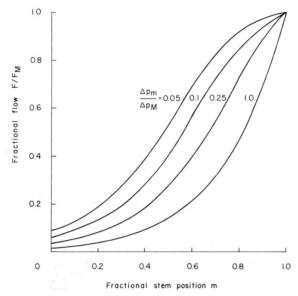

Fig. 1.12. An equal-percentage valve with a rangeability of 50 can provide a nearly linear installed characteristic when allotted as little as 5% of the total system pressure drop.

Even those situations where a pump has been oversized (allowing for too much drop across the control valve) are not beyond correction. If the valve is large enough, as indicated by excessive throttling even at full flow, the pump impeller may be shaved to the point where the valve operates closer to fully open. If necessary, a larger-size valve can be installed, with the characteristic changed from linear to equal percentage.

Quite often, an *installed* equal-percentage characteristic is desirable because of certain nonlinear properties of the process (6). In this event, the addition of an external characterizer may be necessary, using either an arbitrary function generator, or a multiplier or divider as described in Ref. (7).

Some work is proceeding in the area of using power-recovery turbines as control valves (8). While this is a commendable goal, it would seem to be beyond the sphere of control valve manufacturers. Instead, the process designer should specify the power-recovery equipment needed to fit his plant, and then let control valves be selected to regulate that equipment as efficiently as practicable.

3. Minimizing Throttling

When a valve is closed, it develops no power loss because flow is zero. And at full flow, its power loss is low because pressure drop is at its minimum.

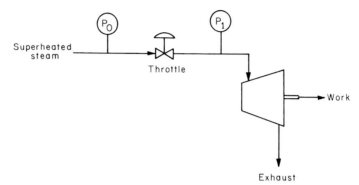

Fig. 1.13.　The throttle valve reduces the available work delivered to the turbine.

Valve losses will be found to pass through a maximum generally between 50 and 75% of full flow.

As an example, consider the high-pressure steam turbine with its throttle valve shown in Fig. 1.13. In general, the turbine-inlet pressure p_1 is directly proportional to steam flow, while the throttle pressure p_0 is held constant by the boiler control system. Typically, for throttle conditions of 2400 psia and 1000°F in a central-station power plant, the turbine-inlet pressure at full rated flow would be about 1600 psia. Table 1.8 shows the entropy increasing at the turbine inlet as the fractional steam flow is reduced by closing the throttle. When the loss in available work per pound of flowing steam $T_0 \Delta S$ is multiplied by the fractional flow F/F_M, the result is the work lost referred to the flow F_M at full load. If multiplied by the rated full load of the plant in pounds per hour, the lost power in Btu/per hour can be estimated. Observe that the lost power passes through a maximum in the vicinity of 60% load.

A similar relationship may be prepared for liquid-phase systems using the equations already developed. A rearrangement of Eq. (1.47) can provide a solution for Δp which, when multiplied by F, gives lost power. This lost

Table 1.8

Loss in Available Work in Steam as a Function of Flow to the Turbine

F/F_M	p_1 (psia)	S_1 (Btu/lb-°R)	$T_0 \Delta S$ (Btu/lb)	$T_0 \Delta S \, F/F_M$ (Btu/lb)
1.00	1600	1.574	20.3	20.3
0.80	1280	1.597	32.2	25.8
0.60	960	1.627	47.8	28.7
0.40	640	1.669	69.7	27.9

power can be normalized by dividing by maximum flow and pressure drop
to give

$$F \, \Delta p / F_M \, \Delta p_M = (F/F_M)[1 - (1 - \Delta p_m / \Delta p_M)(F/F_M)^2] \qquad (1.52)$$

The solution to (1.52) is plotted in Fig. 1.14. Here again, power loss tends to
peak slightly above 50% flow.

To minimize power losses, throttling should be minimized. This requires
that valves be operated in either the closed or fully open position. Since
there can be no production with all the valves closed, a plant should operate
with all valves open. However, this practice would result in loss of all regulation. As a compromise, control schemes should be devised that allow
most valves to operate nearly full open, but with some margin remaining
to exercise control. In the case of the steam turbine, an open throttle at
reduced load requires that boiler pressure *not* be maintained constant, but
vary with the load. This technique, called "sliding-pressure operation,"
can save significant quantities of energy in steam plants, and in fact is applicable to other processes as well. It is already being used in distillation-
column control (9).

In the chapters which follow, each control valve installation is examined
to determine if its position can be maximized without losing control of the
plant. In some cases this can be accomplished within a single control loop,

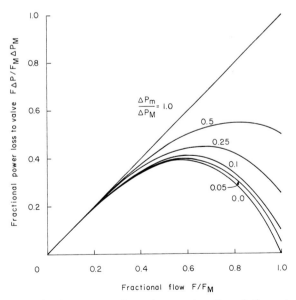

Fig. 1.14. In a liquid-phase system, the peak power loss through the control valve varies
with the pressure allocated to it.

but most applications will require inter-loop coordination. In every case, the requirements of the process for control of quality and quantity must be satisfied first.

REFERENCES

1. Denbigh, K., "The Principles of Chemical Equilibrium," p. 116. Cambridge Univ. Press, London and New York, 1961.
2. Shinskey, F. G., "Distillation Control: for Productivity and Energy Conservation." McGraw-Hill, New York, 1976.
3. Bojnowski, J. H., J. W. Crandall, and R. M. Hoffman, Modernized separation system saves more than energy, *Chem. Eng. Progr.* (Oct. 1975).
4. Johnson, A. J., and G. H. Auth, "Fuels and Combustion Handbook," p. 286. McGraw-Hill, New York, 1951.
5. Sternlicht, B., Low-level heat recovery takes on added meaning as fuel costs justify investment, *Power* (April 1975).
6. Shinskey, F. G., Controlling variable-period once-through processes, *Instrum. Contr. Syst.* (Nov. 1972).
7. Shinskey, F. G., When you have the wrong valve characteristic, *Instrum. Contr. Syst.* (Oct. 1971).
8. O'Connor, J., and H. Illing, The turbine control valve . . . a new approach to high-loss applications, *Instrum. Technol.* (Dec. 1973).
9. Fauth, C. J., and F. G. Shinskey, Advanced control of distillation columns, *Chem. Eng. Progr.* (June 1975).

Part
II

ENERGY CONVERSION PROCESSES

Chapter

2

Combustion Control Systems

Much of the fuel that is consumed in industrial boilers, furnaces, and kilns is wasted because the combustion process is not carefully controlled. Typically, much more than the minimum flow of air is used, to avoid hazards and smoke emission. While this mode of operation is safe, it may be far from economical. The control problem is further complicated when waste fuels of varying composition are introduced for disposal and to recover their heating value.

Air pollution control is another important aspect of combustion. Some undesirable products of combustion, such as sulfur dioxide, cannot be avoided and must be removed external to the combustion process. Others, such as unburned hydrocarbons, carbon monoxide, and the oxides of nitrogen, can be controlled by proper adjustment to fuel–air ratio and flame temperature. The intricate relationships among all these factors are discussed below, leading to a development of control systems which can optimize the performance of the combustion process.

A. CONTROLLING FUEL FLOW

The useful output of the combustion process is heat resulting from the oxidation of fuel. Hence the first item of importance in achieving control over the energy user, whether it be a boiler, gas turbine, or kiln, is the accurate manipulation of fuel flow.

Naturally, each fuel has its own particular characteristics, related not only to its physical state but also to its chemical composition. Accurate metering on a heat-flow basis depends on the physical properties of the fuels at flowing conditions, and their chemical makeup. These properties are evaluated for gaseous, liquid, and solid fuels, as they apply.

1. Gaseous Fuels

The most common gaseous fuel is natural gas, containing typically 75–95% by volume of methane. Other constituents include ethane, propane, and butanes in order of decreasing concentrations, with butanes representing less than 0.5% by volume. The heavier hydrocarbons tend to be more valuable as petrochemical feedstocks, and are therefore being increasingly extracted from the gas prior to distribution. Contaminants include typically less than 1% carbon dioxide, and variable amounts of nitrogen.

Methane oxidizes to form one mole of carbon dioxide and two of water:

$$CH_4 + 2O_2 \rightarrow CO_2 + 2H_2O \qquad (2.1)$$

The heating value or heat of combustion of any fuel yielding water as a product may be expressed in two ways. The higher heating value is obtained by assuming that the water is ultimately condensed, in which case it gives up its latent heat of vaporization. The lower heating value assumes that the water leaves the process in the vapor phase. In both cases, the temperature of the products of combustion is assumed to be the same as that of the reactants.

The higher heating value for methane is given as 1013 Btu/ft^3 at standard conditions, and the lower value as 913 Btu/ft^3. Many natural gases have higher heating values between 1050 and 1090 Btu/ft^3 owing to the presence of heavier hydrocarbons, which more than offset token dilution by carbon dioxide and nitrogen. However, as more of these hydrocarbons are being extracted at the source, heating values continue to fall, approaching 975 Btu/ft^3. Propane is sometimes added on a controlled basis to enrich a poor fuel, and air is occasionally injected to reduce heating value to a specification such as 975 Btu/ft^3.

If the heating value of the gas is controlled by the distributor, then a mass flow measurement is essentially a heat flow measurement. Orifice plates are customarily used for metering natural gas, with mass flow M being related to measured orifice differential h and density ρ:

$$M = k\sqrt{h\rho} \qquad (2.2)$$

(Although a burner tip may be used as an orifice, in which case backpressure is essentially h, variability caused by wear, plugging, etc., advise against this practice.)

Fuel gas is typically accounted in units of standard volume rather than mass. The two are directly related through specific gravity, however. One pound-mole of air occupies 359 ft^3 at standard conditions (60°F and 14.7 psia). The molecular weight of air is taken as 29, and the specific gravity G of a gas is defined as the ratio of its molecular weight to that of air. Then the standard volumetric flow of a gas may be determined from its mass flow:

$$V_s = 359M/29G = 12.38M/G \tag{2.3}$$

The density ρ of an ideal gas varies with its specific gravity, temperature, and pressure:

$$\rho = 29Gp_a/RT_a \tag{2.4}$$

where T_a and p_a are absolute temperature and pressure and R is the universal gas constant in appropriate units. For temperature in degrees Rankine, pressure in psia, and density in pounds per cubic foot, $R = 10.73$.

When (2.3) and (2.4) are combined with (2.2), we have

$$V_s = 20.35k\sqrt{hp_a/GT_a} \tag{2.5}$$

In actual practice, compensation for variations in absolute temperature is not applied as per Eq. (2.5) because flowing temperatures are elevated well above absolute zero. Then the square root of absolute temperature is approximated as a linear function of measured temperature T:

$$\sqrt{T_a} \approx a + bT \tag{2.6}$$

Over the range of 0–100°F, the above approximation is accurate to $\pm 0.1\%$.

For gas pressures below 30 psia, the effect of a variable barometer ought to be taken into account, by using an absolute pressure transmitter. At higher pressures, a gage transmitter may be used, with the measurement elevated in the calculation to a standard atmosphere. At still higher pressures, the supercompressibility factor Z may become significant. Then density varies inversely with Z. For moderate ranges of temperature and pressure, supercompressibility correction can be applied through the linear approximation

$$p_a/Z = f + gp \tag{2.7}$$

where p is the measured pressure and f and g are constants selected for best fit.

Incorporation of the heating value can convert (2.2) or (2.5) into a heat-flow equation; the heating value must be stated in appropriate units, i.e., mass or standard volume. For gas streams, volumetric units are most common:

$$Q = H_s V_s = H_s k_s \sqrt{hp_a/GT_a} \qquad (2.8)$$

where Q is the heat flow rate, H_s is the heating value in standard volumetric units, and $k_s = 20.35k$ from Eq. (2.5). For a gas having a variable composition, both H_s and G will change. These two terms may be singled out of the above equation for purposes of applying compensation for variable composition— this function is known as the "Wobbe Index":

$$\gamma = H_s/\sqrt{G} \qquad (2.9)$$

Fuel gases in most petroleum refineries contain hydrogen and some heavier hydrocarbons along with methane. The effect of hydrogen on the Wobbe Index differs from that of the heavier components. Table 2.1 summarizes Wobbe Indices for several hydrocarbons and selected mixtures. When the index is plotted against specific gravity for each of these gases as in Fig. 2.1, a smooth monotonic curve is formed if hydrogen is absent. This allows a correction factor to be applied to the gas flowmeter as a function of the specific gravity measurement.

However, the Wobbe Index does not change appreciably as the concentration of hydrogen in a mixture varies. Therefore, when hydrogen is the principal variant in a fuel-gas mixture, no compensation need be applied for changes in specific gravity. If specific gravity compensation is being applied to allow for the presence of higher hydrocarbons, variation in hydrogen content will cause problems. And if nitrogen, carbon monoxide, or carbon dioxide are present, the correlation of Fig. 2.1 breaks down completely, in that these components increase specific gravity while contrib-

Table 2.1

Wobbe Indices of Light Hydrocarbons

Composition (by volume)	Heating value (Btu/ft^3)	Specific gravity	Wobbe Index (Btu/ft^3)
H_2	325	0.069	1237
$(H_2 + CH_4)/2$	669	0.310	1201
CH_4	1013	0.552	1364
$(H_2 + C_2H_6)/2$	1059	0.552	1425
$(H_2 + CH_4 + C_2H_6)/3$	1043	0.552	1405
$(CH_4 + C_2H_6)/2$	1403	0.793	1575
C_2H_6	1792	1.034	1762
C_3H_8	2590	1.517	2103
$n\text{-}C_4H_{10}$	3370	2.000	2383

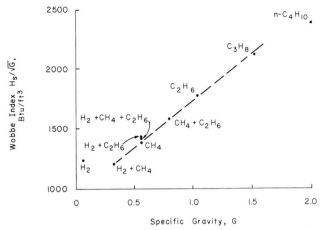

Fig. 2.1. Wobbe Index is related linearly to gas specific gravity for light hydrocarbons and equimolar mixtures of light hydrocarbons with hydrogen.

uting little or nothing to heating value. (The Wobbe Index of carbon monoxide is only 328 Btu/ft^3, while that of carbon dioxide and nitrogen is zero.)

Calorimeters are available to report the heating value of fuel–gas mixtures. These instruments are actually pilot heaters where the fuel is oxidized and the energy released is measured. As such, they are limited by the same type of thermal capacities as any fired furnace, and in general cannot report on rapid changes in heating value in time to be of use to the furnace firing the fuel. Thermal conductivity and density analyzers are not specific, being applicable primarily to binary mixtures such as propane and air. Specific analyzers such as a chromatograph may be used to analyze a fuel gas, but calculating a Wobbe Index from a multicomponent analysis is quite complex, and the chromatograph is not likely to be any faster than a calorimeter.

When the pressure of the fuel gas is variable, a pressure regulator is commonly inserted upstream of the flow control valve, with the flowmeter between them as in Fig. 2.2. Because both valves affect both flow and pressure, interaction may be expected between the two control loops. An analysis made of this process in Ref. (1) indicates that interaction increases proportional to the percentage of total pressure drop taken across the pressure control valve. If, for example, the pressure drop across the pressure control valve is half that across the flow control valve, or less, stable response can be achieved by both controllers. Should the two valves have equal drop, oscillations will break out when both controllers are placed in automatic, if each was adjusted with the other in manual. They then must be readjusted with both in automatic to restore stability.

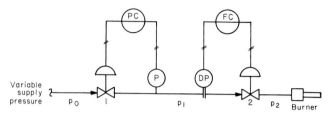

Fig. 2.2. Interaction between the pressure and flow control loops increases with percentage of total pressure drop taken across the pressure control valve.

If the pressure control valve takes a substantially greater drop than the flow control valve, it will have a greater effect on flow than on pressure. As a result, each loop acts as a disturbance to the other, without being able to regulate itself effectively. When both controllers are placed in automatic, flow and pressure are driven away from their respective set points, and regulation is impossible. For the loop assignments shown in Fig. 2.2 to be effective, the pressure control valve should take a smaller drop than the flow control valve, in which case the pressure control valve should be larger than the flow control valve.

An alternate solution to the problem would use only one valve, allowing pressure to go unregulated. In this case, the measured orifice differential would have to be compensated for variations in pressure per Eq. (2.8).

The backpressure developed by the burner varies with flow. As a result the pressure drop across the flow control valve varies inversely with flow. To provide a reasonably linear relationship between fuel flow and valve position, an equal-percentage valve characteristic is desirable.

2. Liquid Fuels

The most common liquid fuels are oils, ranging from light distillate through residual fuel. These oils differ somewhat in heat of combustion and sulfur content, but substantially in viscosity. More important than accurate flowmetering is the proper atomization of oil in the burner. The oil must be available at a controlled pressure and viscosity for complete combustion. This is particularly difficult with heavy residuals in that their viscosity varies sharply with temperature, composition, and even flow. Light distillates may be pumped under controlled pressure and temperature to achieve satisfactory combustion. But a heavy residual must be circulated continuously past the burner and back to a storage tank as shown in Fig. 2.3. If flow is stopped, the oil will cool and plug the line.

In the recirculating system, the net flow of oil to the burner cannot be measured, but must be calculated by subtracting the recirculation rate from the pumped flow. Turbine meters are customarily used for this service

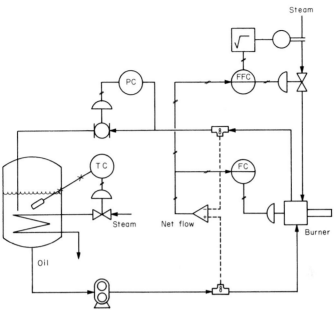

Fig. 2.3. The net flow of oil burned is the difference between the flow to the burner and that recirculated.

because of their superior accuracy, although they are sensitive to viscosity changes, particularly at low flow. Backpressure at the burner is regulated by the recirculation valve, while the firing rate is controlled by adjustment of the burner orifice. Atomizing steam flow should be set in ratio to the firing rate.

If the composition of the oil is constant, temperature control as shown in Fig. 2.3 will generally provide satisfactory regulation over viscosity. However, when oil must be used from several sources, the relationship between viscosity and temperature may be quite variable. In this case, viscosity control is essential, both to provide effective atomization and accurate flowmetering.

A simple viscosity control system is shown in Fig. 2.4. A sample of the heated oil is pumped by positive-displacement action through an orifice, which acts like a model of the burner. Assuming that the flow of the oil is constant, variations in pressure drop across the orifice will indicate variations in oil viscosity and/or density which would have a similar effect on burner performance. Control of orifice pressure drop by manipulation of oil temperature should assure uniform operation of the burner and flowmeters. The sample pump may change its flowrate with time, requiring a readjustment of the viscosity control set point; however, variations in oil density or viscosity cannot affect its pumping rate because they are controlled.

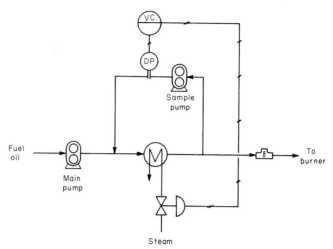

Fig. 2.4. The orifice in the sample line acts as a model of the burner; by controlling its differential pressure for a constant oil flow, burner performance can be regulated.

Oil burner nozzles may be adjusted by a constant-speed bidirectional motor. Oil flow is raised by applying power to the "increase" coil of the motor, and lowered by energizing the "decrease" coil. The flow controller then consists of two comparators, one activated on a positive deviation from set point, the other on a negative deviation. There must be a small dead zone between the two actions to provide stability, and to minimize noise-induced control action. However, the presence of the dead zone limits the accuracy of flow control. To improve the accuracy to which flow may be controlled, the three-state controller described above may receive a burner-nozzle-position signal rather than a flow signal as feedback. Then a conventional proportional-plus-integral flow controller sets the set point of the three-state nozzle position controller in cascade. The arrangement appears in Fig. 2.5.

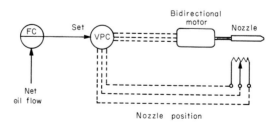

Fig. 2.5. Oil flow can be more precisely controlled if nozzle position is set in cascade by the oil flow controller.

3. Solid Fuels

The most common solid fuel is coal, although waste products such as bark, bagasse, and refuse are appearing as the cost of conventional fuels continues to rise. A solid fuel must be metered, ground, and conveyed to the burner, usually in that order. In a boiler burning pulverized coal, the coal is conveyed to the pulverizer by means of a gravimetric feeder where its mass flow is controlled. The coal is then ground in a mill, with the resulting fine particles being conveyed by a stream of preheated combustion air to the burner. Oversized particles are rejected by a cyclone or screen and returned for regrinding.

Because the flow of coal is controlled at the inlet to the mill rather than at its outlet, there can be some delay in the response of firing rate to a change in coal flow. If the coal is pulverized in a hammer mill, this delay is negligible because there is little capacity in the mill. However, the various types of ball and roller mills may contain from one to several minutes of capacity, in which case the actual flow of pulverized coal lags the feed to the mill by that amount of time.

If the time lag in the mill is relatively short, e.g., a minute or less, it can largely be offset by inserting an equivalent lag in the flow control loop, as shown in Fig. 2.6. This allows the fuel flow controller to have a narrower proportional band than would otherwise be possible, enabling it to largely overcome the lag in the mill. Following an increase in fuel demand, the mill

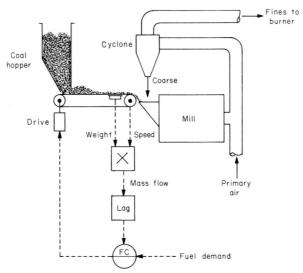

Fig. 2.6. Applying a lag to the coal flow signal gives a dynamic representation of the actual firing rate.

is overfed due to the slow response of the lag unit, until a new steady state is reached. Conversely, a drop in demand will cause a greater dynamic reduction in feed rate, again achieving more rapid boiler response.

When the time lag in the mill is longer than a minute, the flow of primary air, which conveys the pulverized coal to the boiler, must be manipulated to provide response to demands. Unfortunately, however, the amount of coal conveyed per unit of air is not uniform, such that inferring fuel flow from primary airflow alone lacks accuracy. Figure 2.7 shows a means for converting the primary airflow signals to estimated heat release, using measured steam flow. The steam flow signal is combined with the rate of change of boiler pressure to calculate the heat transferred to the steam by the fuel. An integrating controller then applies a correction factor to the total flow of primary air until the estimated heat input agrees with the calculated heat release. The integrator must apply the correction slowly, because it forms a negative feedback loop through the boiler. An increase in steam flow will bring about a decrease in fuel flow through the action of the integrator, which will then cause a decrease in steam flow, tending to form a cycle. However, if the integrator responds more slowly than the boiler, and if pressure is controlled by setting fuel flow, stability can be assured. The time constant τ of the integrator is adjustable in the field. Note that this system also corrects for variations in the heat of combustion of the fuel.

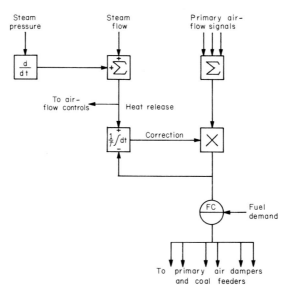

Fig. 2.7.   Steam flow from the boiler can be used to correct primary airflow signals to approximate actual heat released by the fuel.

Lump coal and wood wastes are typically fed onto a traveling grate, where combustion is supported by air directed both over and under the grate. The rate of firing is controlled by the feeder speed. Because the larger particles require more time to be consumed, heat release does not change as quickly with flow as when the fuel is pulverized.

The heat of combustion of solid fuels even from a single source can be quite variable, due principally to moisture content. Coal is stored out of doors, and snow or ice occasionally accompany it into the boiler. Wood wastes may vary in moisture content from a few percent to well over 100% (dry basis), depending on seasoning and environment. Wood should be dried uniformly prior to use as fuel.

4. Coordinating Multiple Fuels

Industry is urged to use waste products as fuel whenever possible, both to reclaim their heating value and to eliminate the pollution that they otherwise might cause. In many cases, if not most, these products tend to be poor fuels with low heating value, which is why they are not salable. In a petroleum refinery, hydrogen and ethane available as byproducts from reforming and cracking operations are usually dumped into a general fuel-gas header. Hot carbon monoxide from catalyst regeneration is usually fed into a separate boiler dedicated to that purpose; supplemental firing with gas may be necessary due to the low heating value of the fuel, particularly if there has already been substantial conversion to carbon dioxide. In a steel mill, blast-furnace gas is usually available, which also has a low heating value, and requires supplementary firing.

Mixing these waste gases with natural gas, as done in a refinery, is not recommended as it produces a fuel of variable heating value and air requirements. Instead, the waste fuel should be fired separately, both to provide more uniform operation and control of the boiler, and also to minimize the use of supplemental fuel. The waste fuel should be fired to the maximum of its availability before calling for supplemental fuel, although some supplemental fuel may be necessary if the heating value of the waste is so low as to cause flameout. It is also important, however, to insure uniform response of the heater control system regardless of which fuel is being fired, or in what combination. Also, limitations on availability of the waste fuel must be enforced. The control system shown in Fig. 2.8 is devised to take all these factors into account.

The signal representing demand for heat is passed through a ratio station where the operator sets K, the maximum allowable percentage of waste fuel to be burned. If K is set at 100%, and the waste fuel is not otherwise limited, waste-fuel flow will equal the demand for heat, such that the two signals to

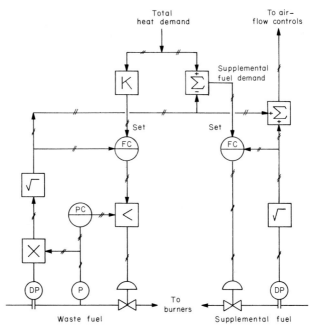

Fig. 2.8. This system allows adjustment over the maximum percentage of waste fuel, accommodates a limit on its availability, and provides uniform response to the demand for heat, regardless of the percentage of waste being burned.

the subtractor will be equal. Then the supplemental-fuel demand is zero. But if K were to be set at 80%, then waste-fuel flow would be 20% less than the total demand, and the subtractor would pass their difference to the supplemental-fuel flow controller. The subtractor must be scaled with a gain equal to the ratio of the full-scale flow of the waste-fuel flowmeter to that of the supplemental fuel flowmeter. Furthermore, the heat-flow range of the total heat demand signal must be identical to that of the waste-fuel flowmeter.

Should the waste-fuel flow range only represent a portion of the total heat demand, additional scaling is necessary. Then K becomes the product of the percentage waste-fuel firing and the ratio of total heat-flow range to waste-fuel heat-flow range. In addition, the negative input to the subtractor must be multiplied by the ratio of waste-fuel heat-flow range to supplemental-fuel heat-flow range; the positive input is multiplied by the ratio of total heat-flow range to supplemental-fuel heat-flow range. This last factor would be unity if the supplemental fuel could carry the full heating load.

Limitation on waste-fuel availability is imposed by the header pressure controller. If there is an excess of waste fuel, pressure will be high and the output of the pressure controller rejected by the low selector. But should header pressure fall below set point, the pressure controller will reduce the waste flow below the flow controller's set point. This change in waste flow is then converted directly into an equivalent increase in supplemental-fuel flow, so that the heater is not thereby upset. Because waste-fuel pressure may vary, its flowmeter must be pressure compensated.

B. CONTROLLING AIRFLOW

Controlling airflow is every bit as important as controlling fuel flow. For complete combustion to take place, there must be an excess of air in contact with the fuel at all times. In the past, air was supplied in plentiful excess, to be certain that all the fuel would be consumed, and therefore smoke and the danger of explosion avoided. However, excess air dilutes the combustion process, reducing flame temperature and increasing stack losses. The heat lost to the stack as a function of stack temperature and percent oxygen in the flue gas is readily calculated, and it demonstrates the need to minimize excess air.

With increasing fuel costs, there is now ample incentive to control excess air precisely. Additionally, as waste fuels with variable heating values are burned, control over excess air becomes more demanding. Furthermore, protection against component failures must be incorporated into these systems. Finally, furnace pressure must be controlled in a safe and stable manner.

1. Combustion Air Requirements

Equation (2.1) described the combustion of methane in oxygen to form carbon dioxide and water. However, most combustion processes use air rather than pure oxygen as an oxidant; the nitrogen passes through essentially unchanged, although its heat capacity has a profound effect on flame temperature. At this point, the combustion reaction is rewritten to include nitrogen, in order to arrive at a molar balance:

$$CH_4 + 2(1 + x)O_2 + 7.52(1 + x)N_2 \rightarrow CO_2 + 2H_2O + 2xO_2 + 7.52(1 + x)N_2 \quad (2.10)$$

Here x is the mole fraction excess air, and 7.52 is the number of moles of nitrogen accompanying 2 moles of oxygen in air.

There are produced 10.52 moles of flue gas for each mole of fuel burned at zero excess air. Excess air adds $2x$ moles of oxygen and $7.52x$ moles of nitrogen. The mole fraction y of oxygen in the flue gas can then be related to excess air as

$$y = 2x/(10.52 + 9.52x) = x/(5.26 + 4.76x) \quad (2.11)$$

The heat leaving the stack per mole of fuel is

$$Q_s = C_p(10.52 + 9.52x)(T_s - T_0) + 34{,}934 \qquad (2.12)$$

where C_p is the molar average heat capacity of the combustion products, T_s the stack temperature, and T_0 the ambient temperature. The constant 34,934 Btu is the latent heat of vaporization of the water formed in combustion. The fractional increase in sensible heat loss per unit excess air near zero excess air can be estimated by differentiating (2.12) and dividing the result by (2.12) evaluated at $x = 0$ less the latent heat contribution:

$$dQ_s/Q_s\,dx = 9.52/10.52 = 0.905$$

In essence, each percent excess air increases stack losses by nearly 1%.

Liquid hydrocarbon fuels such as oils can be represented as a polymer with a formula $(CH_2)_n$, with n being a function of the molecular weight and therefore the volatility of the fuel. A general combustion equation may be written around the CH_2 group alone:

$$CH_2 + 1.5(1 + x)O_2 + 5.64(1 + x)N_2 \rightarrow CO_2 + H_2O + 1.5xO_2 + 5.64(1 + x)N_2 \qquad (2.13)$$

The relationship between flue gas oxygen content and excess air differs somewhat from that calculated for methane:

$$y = 1.5x/(7.64 + 7.14x) = x/(5.09 + 4.76x) \qquad (2.14)$$

Similar relationships can be drawn for other fuels. Rather than develop the combustion reactions for each, the appropriate constants are summarized in Table 2.2 for several common fuels. The relationship between flue-gas oxygen content and excess air all follow the form

$$y = x/(\alpha + 4.76x) \qquad (2.15)$$

Table 2.2

Combustion Air Requirements for Common Fuels

Fuel	Formula	Mole ratio O_2/fuel	Moles O_2 per 10^6 Btu	α
Methane	CH_4	2.0	5.50	5.26
Ethane	C_2H_6	3.5	5.44	5.19
Propane	C_3H_8	5.0	5.38	5.16
n-Butane	C_4H_{10}	6.5	5.37	5.15
Light oil	$(CH_2)_n$	$1.5n$	5.37	5.09
Heavy oil	$(CH)_n$	$1.25n$	5.49	4.96
Hydrogen	H_2	0.5	4.29	5.76
Carbon monoxide	CO	0.5	4.25	5.76
Carbon	C	1.0	5.91	4.76

(Observe that 4.76 is the reciprocal of the oxygen concentration in air.) Table 2.2 includes coefficient α for each fuel.

These fuels are remarkably consistent in their air requirements per Btu and in the relationship between flue-gas oxygen and excess air. This allows airflow to be related directly to heat-flow demand, and excess air to be controlled from a measurement of oxygen in the flue gas, almost without regard to the fuel being burned. Variable mixtures of light hydrocarbon fuel gases show almost no variation in these parameters, and contamination with small amounts of hydrogen and carbon monoxide also have little effect.

Fuel oil is a mixture of liquid hydrocarbons of various molecular weights and heating values, yet their individual combustion constants are so similar that the values shown in the table are quite representative of all. Coal is not pure carbon but is principally a polymer of cyclic aromatic hydrocarbons. The actual parameters for coal will lie between those of carbon and heavy oil, but closer to carbon because its hydrogen/carbon ratio tends to be less than 0.5. Ash in the coal does not enter into combustion, and therefore has no effect on its air requirements.

Ideally, the excess air supplied to a furnace should be zero. However, combustion never goes to completion, and a slight amount of excess air will shift the equilibrium toward lower concentrations of unburned fuel. This is particularly important at low firing rates, where reduced velocities result in poorer mixing of fuel and air. More is said about these matters under the heading of air-pollution control.

2. Fuel–Air Ratio Control

To maintain optimum combustion conditions, airflow should be set in ratio to the fuel flow, if fuel composition is constant. In cases where fuel composition and heating value are not constant, then airflow should be set in ratio to the heat-flow demand. The control system shown in Fig. 2.8 offers two possibilities for setting airflow. A total heat demand signal is available, as well as the output of a summing device which adds the two fuel flow rates. If the air requirements per unit of heat delivered are essentially the same for both fuels, the total heat demand signal may be used. However, if they are different, the summing device may be calibrated in terms of combustion air requirements.

In the event that the heat of combustion or flow rate of the fuel can vary unpredictably, as frequently happens when feeding coal, wood, or refuse, airflow should be set in ratio to the heat released by the fuel. For a boiler, heat release is indicated by steam flow combined with the rate of rise of steam pressure. This is the same calculation that was made in Fig. 2.7 to recalibrate the coal flow measurement. In the steady state, the heat release, coal flow, and total heat demand signal are identical, so any of the three may be used to set airflow. The fastest response to a change in demand will be achieved if total heat demand sets airflow. The best response to an unmeasured change in coal flow or heating value would be obtained if airflow is set by calculated heat release. If the higher of the two signals is selected to set airflow, then excess air will be provided in either event.

The required increase in fuel–air ratio at low loads can be obtained by applying a special characterization to the airflow signal. Airflow is usually measured by a head-type device such as an orifice or pitot tube. To obtain a linear flow signal, the square root of the differential pressure must be extracted. A special calibration may be applied to the square-root function to indicate a lower airflow than what is actually being measured at low loads. The airflow controller then increases the flow, thus providing the desired fuel–air ratio. Figure 2.9 shows a typical curve of percent oxygen desired in the flue gas versus load. Points taken from this curve were used to develop the modified calibration for a square-root extractor for an oil-fired furnace. The calculations are summarized in Table 2.3.

Since both fuel and airflow are set equal to heat demand, the output of the square-root extractor is equal to the heat load appearing in the first column. The flowmeter differential pressure required to produce the excess air required at each load appears in the last column.

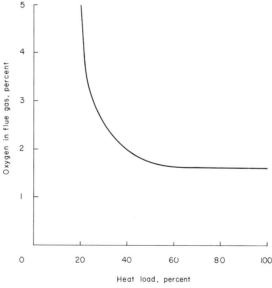

Fig. 2.9.   Flue-gas oxygen content needed for efficient combustion as a function of heat load. (From B. Lipták, "Instrument Engineers Handbook," Vol. II. Chilton, Philadelphia, 1970.)

Table 2.3

Airflow Differential Required to
Produce the Flue-Gas Oxygen Program of Fig. 2.9

Heat load (%)	O_2 in flue gas y (%)	Excess air x (%)	Unadjusted differential (%)	Adjusted differential (%)
100	1.6	8.7	100.0	100.0
70	1.6	8.7	49.0	49.0
50	1.7	9.3	25.0	25.3
40	2.0	11.1	16.0	16.7
25	3.0	17.5	6.3	7.3
20	5.0	32.8	4.0	6.0

From Table 2.3 it can be seen that very little error in the airflow signal or the fuel flow signal can have a substantial effect on flue-gas oxygen content. To maintain the optimum quantity of excess air, the fuel–air ratio should be automatically adjusted by a controller acting on the measured amount of oxygen in the flue gas. Because the correct air–fuel ratio can be predicted within relatively narrow limits, the range of the adjustment can be restricted. Figure 2.10 outlines an airflow control system including feedback trim of oxygen content. The set point for the flue-gas oxygen controller must be characterized as a function of steam flow in accordance with Fig. 2.9.

There are various types of oxygen analyzers for use in flue gas. Sampling is difficult because of high temperatures, corrosive conditions, and the presence of ash and soot. An in-situ measurement is least likely to be affected by leaks, condensation, etc. However, there is a risk of failure, and the control system must be protected. Limits on the output of the oxygen controller can

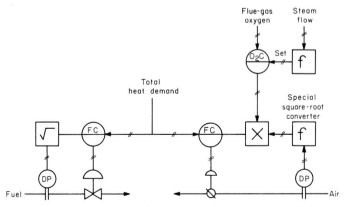

Fig. 2.10. The flue-gas oxygen controller automatically corrects the air–fuel ratio for errors in the metering system.

keep the air–fuel ratio in reasonable bounds in the event the measurement should fail without notice.

A probe-type oxygen analyzer which inserts directly in the flue-gas duct produces a voltage which varies with the logarithm of the oxygen content (2). Because the sensitivity of the signal to oxygen content varies inversely with the oxygen content, the gain of the oxygen control loop could be expected to change with load. At minimum load, the required oxygen content may be three times as great as at full load, so the sensitivity of the analyzer is only one-third that at full load. However, the delay in response of oxygen content to a change in airflow is longer at low load than at high load. This increasing delay increases the gain of the integral controller used on the oxygen signal, which virtually offsets the reduced gain of the logarithmic signal. Consequently, the logarithmic analyzer seems well suited to this application. In addition, the function given in Fig. 2.9 is less extreme when characterized to the nonlinear O_2 scale.

Gilbert (2) claims superior control of fuel–air ratio can be achieved by controlling CO emission in the 100–200-ppm range. His reasoning is that air infiltration or a maladjusted burner could give a satisfactorily high O_2 level while still not providing complete combustion. On the other hand, a CO measurement is not affected by air infiltration, and is a true indication of the completeness of combustion.

3. Maintaining Safe Conditions

Because of the severe damage possible should an explosive mixture of fuel and air accumulate in a hot furnace, an elaborate system of interlocks is justified. Before fuel can be admitted, the firebox must be purged with air sufficiently long to ensure that fuel will not be present in an explosive concentration. Then a pilot flame must be ignited and verified by a flame detector before the main fuel valve is allowed to open.

The furnace must also be protected against an excess of fuel, in the event of a control or equipment failure. This protection is provided by the pair of signal selectors shown in Fig. 2.11. Should airflow fail to respond to an increase in the demand for heat, the low selector will set fuel flow equal to airflow. Should fuel fail to respond to a decrease in demand for heat, the high selector will set airflow equal to fuel flow. Thus air is forced to lead fuel on an increasing load, and fuel to lead air on a decreasing load. This arrangement also protects against a fan failure or a sticking fuel valve.

For the dual selector system to function, air and fuel flows must be scaled on the same heat-equivalent basis. Here, volumetric or mass flow units are meaningless. This is why recalibration of the air system is provided on the measurement side of its flow controller rather than on the set-point side.

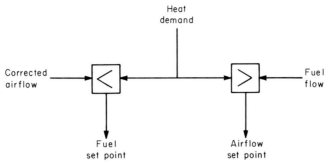

Fig. 2.11. The signal selectors prevent an excess of fuel from developing during changing load or upon a fan failure.

The high selector could receive a third input from the heat release calculator in the event that recalibration is also supplied to the fuel flow measurement, as in Fig. 2.7. In absence of fuel flow recalibration, however, the heat release signal will not likely agree with fuel flow, and the dual selector system will then fail to function properly.

The dual selector system forces airflow to lead fuel on an increasing load and to lag on a decreasing load. Then flue-gas oxygen content tends to deviate above the set point on *all* load changes. If the oxygen controller were allowed to react proportionally to these deviations, it would tend to defeat the security provided by the selectors. Consequently, the integral control mode alone is used on the oxygen signal, so that reaction to rapid fluctuations is minimized. The principal function of the controller is to correct for long-term deviations caused by flowmeter errors and variations in fuel quality.

4. Airflow and Furnace Pressure Control

Typically oil and gas are fired under a slight positive pressure—10–20 in. of water—supplied by a forced-draft fan. If incoming air is preheated against flue gas, or if there is pollution-control equipment such as precipitator, filter, or scrubber on the flue gas, an induced-draft fan is also required. Coal and wood are fired under a slight negative pressure, less than an inch of water, requiring an induced-draft fan at the stack. Again, if combustion air is preheated, a forced-draft fan is usually required. So most modern furnaces of substantial size will have both types of fans, with dampers arranged as shown in Fig. 2.12.

Although the figure shows airflow being measured by an orifice, this additional restriction is often avoided by measuring the differential pressure between the windbox and the furnace, or across the furnace. Temperature

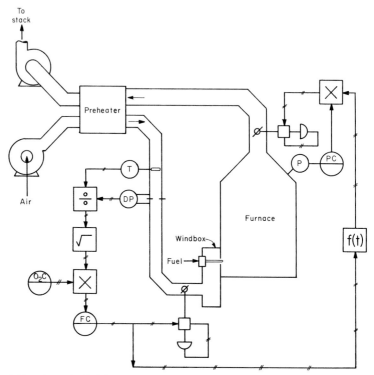

Fig. 2.12. Control of airflow and furnace pressure requires the coordination of inlet and outlet dampers.

compensation is usually needed for preheated air, since its temperature will vary with both the heat load and the ambient.

The relationship between the mass flow of air M and measured differential h across a restriction is a function of absolute temperature T:

$$M = k\sqrt{h/T} \qquad (2.16)$$

where k is a proportionality constant. To find the error caused by a temperature variation, M is differentiated with respect to T:

$$dM/dT = -k\sqrt{h/T}\,T\sqrt{T} \qquad (2.17)$$

Substituting (2.16) into (2.17) allows the expression of flow error dM relative to flow M:

$$dM/M = -dT/T \qquad (2.18)$$

If temperature varies 10°F in 240°F (700°R), the relative flow error dM/M would then be 10/700 or about 1.4%.

Temperature compensation can be readily applied using Eq. (2.16), using a convenient range of measurement such as 0–300°F. The signal may be given the necessary elevation to absolute temperature by scaling the divider appropriately. Equation (2.19) illustrates the scaling for the divider:

$$M' = \sqrt{\frac{h'(460 + T_b)}{460 + T_0 + (T_M - T_0)T'}} \qquad (2.19)$$

where M' is the mass flow signal (0–100%), h' the differential pressure signal (0–100%), T_b the base temperature (°F), T_0 the temperature at zero scale, T_M the temperature at full scale, and T' is the temperature signal. For the example where temperature is measured over the range of 0–300°F, and the flowmeter is selected for a base temperature of 200°F,

$$M' = \sqrt{h'/(0.697 + 0.455T')}$$

In cases where combustion air is split between primary and secondary flows for purposes of coal feeding, etc., the various flow measurements must be linearized before being added. Temperature compensation may then be applied to the summed signals without the need for linearizing the temperature signal. Because measured temperature is elevated well above absolute zero, there is little error involved in a linear approximation of the square-root curve over the operating range. Compensation is applied per Eq. (2.20):

$$M' = \sum \sqrt{h'/(a + bT')} \qquad (2.20)$$

where a and b are scaling coefficients. Equations (2.19) and (2.20) may be equated at T_0 and T_M by setting T' at zero and unity. There will then be no error due to linearization at those two points. Then

$$a = \sqrt{(460 + T_0)/(460 + T_b)} \qquad (2.21)$$

$$a + b = \sqrt{(460 + T_M)/(460 + T_b)} \qquad (2.22)$$

For the same conditions given in the above example, $a = 0.834$ and $b = 0.240$. The approximation in this example yields a maximum error of -0.8% at midscale temperature. If coefficient a is raised to 0.838, the error is distributed over the temperature range to $\pm 0.4\%$.

A natural interaction exists between airflow and furnace pressure similar to that encountered between fuel-gas flow and pressure. In the airflow system, however, the two dampers are generally the same size and carry the same pressure drop. In this case, interaction exists which cannot be overcome by changing control-loop assignments. Instead, decoupling is recommended as shown in Fig. 2.12.

With this system, both dampers are manipulated to control airflow. If the dampers are properly characterized and matched, moving them equally should not have a residual effect on furnace pressure. However, the furnace-pressure controller is capable of moving the downstream damper to correct for any deviation which might develop due to a mismatch. Depending on the location of the pressure measurement, it may respond more rapidly to the downstream damper than it does to the upstream damper. Then a dynamic lag $f(t)$ should be applied to the signal from the first to the second damper to match their dynamic effect on pressure.

For simplicity, there is no decoupler shown to prevent the furnace-pressure controller from upsetting airflow. If the decoupler shown is effective, very little corrective action will be required of the pressure controller, and the interaction between the two loops will not be significant. Furthermore, the flow controller must respond to all load changes, whereas the decoupler removes that requirement from the pressure controller.

C. AIR POLLUTION CONTROL

The degree of air pollution caused by the combustion of a fuel is very dependent on the type of fuel. Hydrogen would seem to be ideal in this respect, in that it can be obtained in rather pure form, i.e., free of contaminants such as sulfur, and burns to completion yielding only water. However, the formation of oxides of nitrogen will take place to some extent, at combustion temperatures. Natural gas can produce some carbon monoxide and possibly unburned fuel as well. Oil and coal both contain sulfur which oxidizes to sulfur dioxide and trioxide. Furthermore, coal contains some nonvolatile materials which may be carried out the stack. Each of these pollutants requires control related to its particular characteristics.

1. Carbon Monoxide and Unburned Hydrocarbons

These two pollutants are lumped under a single heading because they have the same cause—incomplete combustion. Complete combustion depends on two factors: excess air, and mixing of fuel and air. Gaseous fuels mix easily with air and therefore tend to burn completely with relatively little excess oxygen, perhaps 1.5%. Liquid fuels do not mix as well, particularly those with high viscosity, requiring high-pressure injection with steam atomization for efficient firing. Even then, more excess air is required for complete combustion than with gases. Solids are still more difficult to mix with combustion air, and consequently require still more excess air.

An equilibrium naturally exists in any combustion process between the hydrocarbon fuel, carbon monoxide, and oxygen:

$$[CH_2] + 1.5O_2 \rightleftharpoons H_2O + CO + 0.5O_2 \rightleftharpoons H_2O + CO_2 \qquad (2.23)$$

Oxygen drives the equilibrium to the right. In a well-mixed system with excess air, the concentration of unburned hydrocarbons $[CH_2]$ must be less than that of CO. Therefore, if the CO content is controlled to a satisfactory degree, the concentration of hydrocarbons must necessarily be low. If the reasoning is carried further, then control of flue-gas oxygen content implies control of CO in equilibrium with it.

However, the equilibrium depends to a great extent on mixing. A deficiency of air at one location in the combustion zone could result in release of unburned hydrocarbons and CO, even with O_2 in the flue gas. Inefficient atomization due to a fouled burner nozzle could have a similar effect. And, naturally, mixing is less thorough at low loads where both fuel and air velocities are reduced.

Consequently, control over oxygen content does not necessarily ensure control over CO and hydrocarbon emissions, even when properly characterized with load. May (3) reports that maximum combustion efficiency can be obtained with 1–2% oxygen when burning natural gas, and 2.5–3.5 % with No. 6 fuel oil. If maximum efficiency were to be achieved at still higher levels, this indicates that burners are in need of cleaning or adjustment.

Carbon monoxide in the flue gas can be measured with an infrared analyzer. If the air–fuel ratio were to be adjusted to maintain a constant CO level in the flue gas, the resulting oxygen content would be indicative of the condition of the burners at any given load.

2. Oxides of Nitrogen

At high temperatures ($> 3000°F$), flue-gas molecules begin to dissociate into ions. An equilibrium then takes place wherein the ions of nitrogen and oxygen may combine to form nitrogen oxides:

$$[N] + x[O] \rightleftharpoons NO_x \qquad (2.24)$$

The oxides are a mixture of NO, NO_2, N_2O_4, and N_2O_5, which will combine with moisture and air to form nitric acid.

NO_x formation is promoted by elevated temperature and pressure, and excess air. Engines tend to generate considerable NO_x because combustion takes place under pressure. Bell and Breen (4) report that combustion of gas typically produces 100 ppm of NO_x, while light distillate yields 150 ppm, residual oil 250 ppm, and coal as high as 500 ppm. This relationship is undoubtedly due to the excess oxygen typically required for combustion of these fuels. If the predominant oxide formed is NO_2, the rate of oxide formation will be proportional to the square of the oxygen content of the flue gas. If coal typically requires twice the excess oxygen for efficient combustion as gas, it would be expected to produce four times the NO_x concentration.

Any modification or adjustment to the burners which will allow more complete combustion of fuel with less excess air will also reduce NO_x emission. Consequently, burner design and maintenance are most important in reducing this source of pollution. And close control of excess air will have the additional benefit of reducing NO_x emission along with heat loss.

The other controllable parameter contributing to NO_x formation is the combustion temperature. Temperatures in the combustion zone can best be reduced by dilution of fuel with flue gas, most of whose sensible heat has been extracted. It is difficult to measure temperatures in the combustion zone reliably due to variability of the flame pattern, oxidation, erosion, etc. However, flame temperature can be regulated by recirculating a certain portion of flue gas into the combustion air.

If the temperature of the flue gas were constant, a fixed ratio of recirculation to airflow would produce a constant flame temperature. However, flue-gas temperature tends to rise with load, requiring an increasing recirculation ratio. A heat balance on the flue gas indicates that flame temperature is reduced from its maximum value T_M to T by recirculating X fraction of flue gas having a temperature T_w:

$$X = (T_M - T)/(T - T_w) \qquad (2.25)$$

About 25% recirculation of flue gas at 500°F will reduce flame temperature from 3000 to 2500°F.

Figure 2.13 describes a control system setting the flow of recirculated flue gas in proportion to steam flow through a ratio flow controller (FFC). Temperature compensation of the flue-gas measurement is not actually necessary. While Eq. (2.25) requires a higher flow at higher temperature, the flowmeter indicates a lower flow due to the effect of temperature on gas density. The two effects are not identical, but they cancel one another within the limits of flowmeter accuracy.

Increasing the ratio setting will bring about more recirculation, thereby reducing NO_x emissions. However, boiler efficiency tends to be reduced

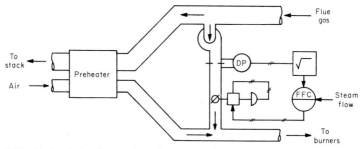

Fig. 2.13.   Recirculating flue gas in ratio to steam flow will regulate flame temperature and thereby reduce NO_x formation.

because the temperature gradient between the flue gas and steam is reduced. Therefore, an optimum ratio should be found which will hold NO_x content at its specified limit and no lower.

Reference (5) indicates flue-gas recirculation can reduce NO_x emissions by 60% for gas firing, 20% for oil firing, but is ineffective with coal firing. Some NO_x will be absorbed along with SO_2 in downstream wet scrubbers.

3.   Particulates

Particulates are problems only when solid fuels are fired. They tend to be hard, sharp, and porous, readily absorbing both acids and moisture. If the flue-gas temperature is allowed to fall below 300°F, acids in the flue gas will condense and cause extensive corrosion. If particles are removed by a bag filter, however, temperatures must be maintained below 450°F to protect most fabric materials. Bag filters are therefore installed downstream of any air preheater, with a bypass around the preheater allowing hot flue gas to control stack temperature above 300°F. Periodic blowback or mechanical agitation removes trapped particles.

Electrostatic precipitators may be used either before or after the air heater, depending on the resistivity of the particles. If the sulfur content of the fuel is high (3% or more), a conductive film of acid is absorbed on the particles, and they conduct readily in the 300–400°F range found downstream of the air heater. For fuels with lower sulfur content, the precipitator is ordinarily installed upstream of the air heater, where temperatures in excess of 450°F ensure adequate conductivity. This has the advantage of keeping particulates out of the air heater. Precipitators are equipped with controls to maintain field charge and avoid corona formation. Periodic rapping of electrodes removes accumulations of trapped particles.

Wet scrubbers are generally used in conjunction with a wet desulfurization system. Particles are collected most effectively when the water stream contacts the flue gas co-currently. Water droplets envelop and agglomerate the solid particles, and are then removed from the gas in a cyclone or baffled separator. Venturi or jet scrubbers are most common for particulate removal.

Devices like venturis and cyclones are sensitive to changes in velocities, complicating their use in flue-gas handling. To accommodate the variable gas flows anticipated in power plant operation, some venturis are equipped with an adjustable throat, as shown in Fig. 2.14. The throat area is automatically adjusted to maintain a constant differential pressure across it, hence a constant velocity.

After particulate material has been removed by co-current high-velocity contact, fumes such as sulfur dioxide are absorbed in countercurrent low-velocity contact with a scrubbing medium. The exiting flue gas becomes cooled and saturated with water in its passage through the scrubbers. In this

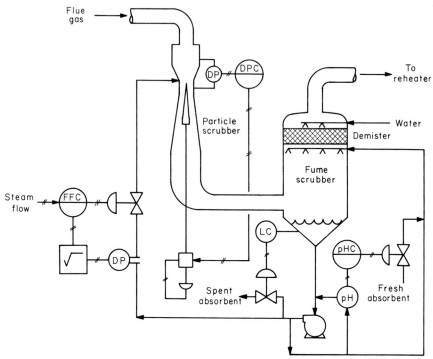

Fig. 2.14. The throat of the venturi is automatically adjusted to maintain a desired gas velocity; circulation is set in ratio to boiler load.

state it will not rise well from the stack, and will develop a noticeable steam plume when it contacts the atmosphere. To avoid these problems, the saturated flue gas is reheated from 180 to perhaps 250–300°F with steam or fuel before entering the stack. Since this reheat energy is a total loss, stack temperature should be held to a minimum consistent with ambient conditions. Stack temperature should be controlled at a specified differential above ambient to provide the density difference required for lift. However, the temperature difference should be adjusted relative to humidity and wind velocity to control plume visibility consistent with minimum reheat.

4. Oxides of Sulfur

With decreasing availability of high-quality fuels, those containing increasing amounts of sulfur are being used. In the process of combustion, the sulfur is oxidized principally to the dioxide, with perhaps 1% formation of trioxide as a function of excess air and residence time in the furnace. When exhausted to the atmosphere, eventually all of the SO_2 will be oxidized to

SO_3, and be returned to the ground as a dilute solution of sulfuric acid in rainwater. Corrosion of metals and concrete from acid rainwater is common in industrialized areas, and the pH of lakes and rivers has also been falling to the detriment of aquatic life.

The oxides of sulfur are quite readily absorbed by alkaline solutions based on sodium, calcium, magnesium, ammonium, or even organic reagents. In most of these processes, the sulfur is converted to a sludge of calcium sulfite and/or sulfate and landfilled. However, the Japanese recover calcium sulfate for wallboard manufacture. Some processes recover SO_2 from the scrubbing medium either by evaporation or by kilning; it is then converted to sulfuric acid or reduced to elemental sulfur.

The variable having the most pronounced effect on the SO_2 content of the effluent is the pH of the scrubbing medium. Absorption is based on the solubility of SO_2 in equilibrium with its partial pressure in the gas phase. To lower that partial pressure, the concentration of dissolved SO_2 must be reduced by conversion to the bisulfite (HSO_3^-) ion or to solid calcium sulfite. In the first reaction, soluble sodium sulfite is used as the scrubbing medium, such that sulfite ions convert SO_2 to the bisulfite:

$$SO_3^{--} + H_2O + SO_2 \rightleftharpoons 2HSO_3^- \tag{2.26}$$

In the second, calcium ions are available from either the hydroxide (lime) or carbonate (limestone):

$$Ca^{++} + SO_2 \rightleftharpoons CaSO_3 + 2H^+ \tag{2.27}$$

In both cases, the absorption of SO_2 causes a reduction in pH, which tends to reduce the rate of absorption.

Figure 2.14 shows a pH controller adding fresh reagent to maintain an effective composition for the scrubbing solution. The optimum pH depends on the nature of the process. When scrubbing with lime or limestone, a pH of about six provides essentially complete neutralization. In fact, it is impossible to raise the pH above about 6.7 with limestone because of its limited solubility.

When scrubbing with a sodium-based reagent, the reaction endpoint occurs closer to pH 5. If too high a pH is used, carbon dioxide absorption increases, using reagent needlessly and forming carbonate scale within the scrubber itself. Sodium-based reagents are usually regenerated by reaction with lime in a second stage. Controls for this operation are described in Ref. (6).

5. Flare-Stack Controls

Flares are common in petroleum refining and chemical production, to incinerate gaseous wastes and products vented through relief devices. If

steam is injected into the flame, radiation and smoke can be substantially
reduced. The difficulty arises that the quantity of gas being flared can vary
over a wide range, and change quite rapidly.

There are two approaches to this problem: steam flow can be set in ratio
to gas flow, or adjusted on the basis of radiation admitted by the flare. The
ratio control (feedforward) system has some drawbacks. Orifice flowmeters
customarily used to measure gas and steam flows are limited in rangeability
to only 4:1, too low for most flare applications. Higher rangeability flowmeters, such as the vortex-shedding type, are quite expensive in the large
sizes needed for most flares. In addition, the composition of the flared gas
may change radically, so that the steam-to-gas ratio should be variable.
To avoid smoke formation when flaring rich mixtures, a high ratio must be
used, which is wasteful of steam when leaner mixtures are flared. Finally, the
steam flow must be made to lead the gas flow, to compensate for the lag in
the flow control loop, and the transportation of the steam from the valve to
the top of the flare.

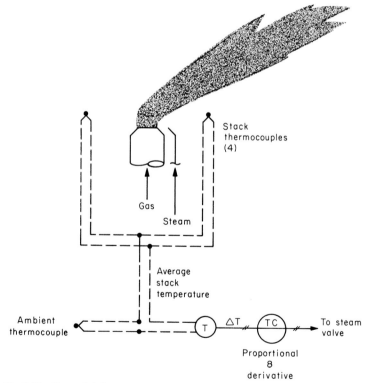

Fig. 2.15. Steam is injected into the flare to reduce radiation and smoke emission.

An alternate system based on feedback (7) is shown in Fig. 2.15. An array of four thermocouples is placed around the exit of the flare stack, exposed to the radiation of the flame. They are wired in parallel, giving an average of the four temperatures. At least one will be exposed to radiation from the flare, regardless of the direction of the wind at any time.

Should there be little flame, temperatures would approach ambient, and the voltage developed against a fifth thermocouple exposed at the base of the stack would approach zero. But when a bright flame begins to develop, a differential temperature will appear between the flare and the ambient. A proportional-plus-derivative controller acting on this temperature rise will open the steam valve. As the radiation is decreased by the steam, the temperature differential will fall, although it cannot be reduced to zero because of the heat released by the flare. Hence the controller cannot have integral action, or it would continue increasing steam flow without being able to further reduce temperature. Derivative action will help overcome the lag in the thermocouples and steam line.

The controller should be biased so that, at zero temperature difference, the steam valve is closed. The proportional band should be set as low a value as possible to minimize smoke; however, if it is set too low, cycling will develop as in any feedback loop. Derivative time should be set to reduce the period of the cycling by about 30%, which will allow a slight further reduction in the proportional band, while improving the speed of response of the loop.

Rangeability is limited to that of the valve, typically 50 to 100 : 1. A linear valve is recommended.

REFERENCES

1. Shinskey, F. G., "Process Control Systems," pp. 188–190. McGraw-Hill, New York, 1967.
2. Gilbert, L. F., Precise combustion-control saves fuel and power, *Chem. Eng.* (June 21, 1976).
3. May, D. L., Cutting boiler fuel costs with combustion controls, *Chem. Eng.* (Dec. 22, 1975).
4. Bell, A. W., and B. P. Breen, Converting gas boilers to oil and coal, *Chem. Eng.* (Apr. 26, 1976).
5. Environmental Protection Agency, Rep. No. 650/2-74-066, U.S. E.P.A., Washington, D. C.
6. Shinskey, F. G., "pH and pIon Control in Process and Waste Streams," pp. 88–92. Wiley (Interscience), New York, 1973.
7. Environmental Protection Agency, Controlled flare cuts smoke, noise, saves steam, *Technol. Transfer*, U.S. E.P.A., Washington, D.C.

Chapter

3

Steam Plant Management

An integral part of every industrial complex is the steam plant or boiler house. In small facilities, the boilers are usually low-pressure units which supply steam for process and space heating only. Mechanical power would then have to be supplied primarily by purchased electricity. However, much energy can be saved if high-pressure steam is generated and used to drive mechanical equipment, with the engine exhausts dedicated to heating service. Therefore, new process plants and sizeable existing units have high-pressure steam available for driving turbines. Yet if an optimum balance exists between work and heat in any of these facilities, it would be pure accident. While a particular plant might once have been designed for that optimum balance, subsequent changes needed for production or associated with haphazard expansion usually leaves the plant far away from it. Ordinarily there is no attempt made to restore the balance when new equipment is purchased.

Yet the principal reason for a lack of balance at any point in time is lack of control. Although the various headers may be held precisely at their specified pressures, and all individual users are under control, they are not necessarily coordinated with each other. Much of this lack of coordination can be overcome with a control system that looks at both users and suppliers, and strives for optimum conditions as opposed to an arbitrary set of conditions.

This entire philosophy is new to industrial boiler plants. In the past, there has been little communication between the steam plant and the users other than complaints when pressure was too low or too many units were drawing steam at the same time.

Recognize that the products from most process plants contain no more energy than the feedstocks, and usually less. In this context, neither the boilers nor the electrical feeders put any energy into the products—all of the energy they supply ultimately only warms the environment. Then considerable energy can be saved by optimizing the way in which it is applied to the process—the multiple-effect evaporator is a notable example.

The incentives for optimum steam plant management are many. Not only will fuel be saved but feedwater as well, and wear in valves, pumps, and turbines will be reduced. Of equal importance, the loading of the steam plant can be substantially reduced so that a production increase could be obtained without adding another boiler.

A. STEAM TURBINE CONTROLS

The real challenge in an industrial steam plant is balancing heat and work. Electricity purchased from a central-station power plant contains typically only 32% of the energy supplied to that plant. Therefore, purchased electricity carries a premium price. Power generated in the industrial steam plant *can* cost considerably less, if the plant is properly balanced.

In general, mechanical equipment is more efficiently driven by steam turbines than by constant-speed electric motors. Although small turbines are not as efficient as the larger ones which are used to drive generators, they avoid the conversion from shaft work to electricity and back. Secondly, steam turbines may be operated at reduced speeds, thereby allowing energy savings in a controlled situation. These drives must be manipulated to satisfy the power requirements of the process, in the same way that its heating demands must be met. If a process's power demands exceed its proportionate heating demands, there will be an excess of exhaust steam which needs to be condensed and therefore wasted. If the heating requirements exceed the need for power, some steam must bypass the turbines to make up the deficiency. This represents lost work and should be minimized.

Flexibility can be added to the plant, to allow heat–work balancing, if one or more turbines is used to generate electricity. Then if an excess of exhaust steam appears, the turbine(s) generating electricity can be cut back, and more electricity will then be drawn from the power lines. If exhaust steam is insufficient to supply all the process heating needs, then electrical generation can be increased to provide the balance. In this case, the flow of purchased power would be reduced, or, in the extreme, the plant could sell power to the utility. While this situation is not commonly encountered, it should be an important consideration for the future.

1.   Condensing and Noncondensing Turbines

Because central-station power plants have no market for low-pressure steam, their turbines are designed to discharge directly into a condenser. Their cycle efficiency is maximized when the condenser temperature and pressure are minimized. Some turbines in industrial boiler houses are also of the condensing type. They are virtually all used for generating electricity, because there is no other demand for shaft work in the boiler house where a large condenser would be located.

Turbines used to drive pumps and compressors are scattered about the process plant. Because of their distribution and wide variety of sizes, it does not make sense to use condensers with these turbines. However, they cannot be allowed to exhaust to the atmosphere, as this would waste valuable feedwater and energy as well. Consequently, these turbines generally exhaust into a header where the pressure is high enough for the steam to be useful for heating.

In this context, noncondensing turbines are frequently called "back-pressure" turbines, as they discharge into a backpressure maintained on the header. The lower that backpressure the more energy can be extracted from the steam by the turbine. Consider, for example, steam at 600 psia and 600°F throttled to 500 psia at the inlet of a backpressure turbine. Steam enthalpy at the inlet is 1290 Btu/lb. If the steam is expanded isentropically to 150 psia, its enthalpy at discharge would be 1188 Btu/lb, the difference of 102 Btu/lb being converted into work. However, turbines are typically only 70% efficient, such that only 71.4 Btu/lb would be converted to work, leaving the exhaust at 1218.6 Btu/lb and slightly superheated instead of wet. If the turbine backpressure could be reduced to 120 psia without changing inlet conditions, isentropic expansion would reduce the enthalpy to 1160 Btu/lb. At 70% efficiency, the enthalpy converted to work would be 91 Btu/lb, an increase in shaft work of 27.5% over operation at 150 psia backpressure.

By contrast, variations in supply pressure have very little effect. If the same turbine is supplied steam at 550 psia and 600°F, and again throttled to 500 psia, exactly the same work is recovered in discharging to 150 psia as before. The throttle must open wider to deliver the same steam flow with half the pressure drop, but that is the only difference. A general conclusion can then be drawn that all header pressures in a steam plant should be as low as possible, consistent with meeting the control objectives of the process.

In general, the pressure at the inlet of a turbine (downstream of the throttle) varies linearly with steam flow. In fact, the inlet pressure to the first stage in large generating plants is used as a flowmeter, being virtually as accurate as a nozzle, and linear, where a nozzle differential pressure is related to the square of flow. Variations in backpressure have relatively little effect on the flow-inlet pressure relationship as long as the inlet pressure is greater than twice the absolute backpressure. Consequently, changing the backpressure in the earlier example did not alter either the flow or inlet pressure.

However, if a turbine delivering shaft work to a pump or compressor sustained a backpressure change, the work delivered to the process would be affected. Then the process controls would have to readjust the throttle to restore the controlled variable to its previous value, producing a net change in steam flow and therefore inlet pressure. For example, the 27.5% increase in shaft work per pound of steam brought about by the reduction in back-pressure from 150 to 120 psia would promote a decrease in steam flow. If the steam flow were decreased by the same proportion, however, the throttle would have to lower the inlet pressure to 392 psia. But the loss in available work caused by this much throttling offsets most of the effect of the reduced backpressure—shaft work would be reduced to 77 Btu/lb. Consequently, the throttle will ultimately seek a middle position, giving an inlet pressure of about 430 psia and a reduction in steam flow of perhaps 14% of the original. Table 3.1 summarizes these conditions.

If the turbine were being used to generate electricity, it could be operated in an unconstrained mode, i.e., accepting all available steam. Then the entire increase in work caused by the reduction in backpressure would have been usable.

In the inflexible case where all turbines are under process control, a heat–work balance can sometimes be reached by adjusting turbine backpressure. When work demands exceed the need for heat, excess low-pressure steam may have to be vented, condensed, or better, used to heat boiler feedwater.

Table 3.1

Effect of a Reduction in Backpressure on a Noncondensing Turbine

Inlet pressure (psia)	Steam flow (lb/hr)	Backpressure (psia)	Enthalpy loss (Btu/lb)	Work (Btu)
500	1.0	150	71.4	71.4
500	1.0	120	91.0	91.0
392	0.78	120	77.0	70.1
430	0.86	120	82.7	71.1

However, if its pressure can be adjusted downward, more work can be extracted per pound of steam, and the turbine controls will then reduce the steam flow accordingly. In many cases, this could reduce the excess flow to zero, and thereby save considerable energy. Naturally, constraints on header pressure exist—if it were lowered too much, some of the users may fail to obtain their required heat. A discussion on header-pressure optimization follows later in the chapter.

2. Extraction Turbines

Most condensing turbines have taps where steam can be extracted between stages for heating. In central-station power plants, steam is extracted at several points for feedwater heating, leading to substantial improvement in efficiency. There are obvious limits to this practice. While increased extraction recovers more heat, it produces less work as well. Furthermore, the extracted steam gives up its latent heat to the sensible heat of the feedwater. Consequently, only a relatively small fraction of the steam may be usefully extracted for feedwater heating. And because the heat transfer is non-isothermal, heating must be broken into several stages, each using steam extracted from a different point in the turbine.

Again, central-station power plants have no other use for extracted steam than to heat feedwater. Industrial steam plants have manifold uses for steam as well as work, and therefore have turbines specifically designed for large extraction flows. Reference (1) describes both single and double extraction turbines which supply large volumes of steam to low-pressure headers. They are designed to operate at throttle pressures to 1250 psia or more, discharging into a condenser. Extraction pressures are typically in the 150–250 psia range for the single-extraction turbine, with a second discharge in the 40–50 psia range for the double-extraction unit.

In essence, the turbines are segregated with throttle valves between the sections as shown in Fig. 3.1. The valves are usually linked mechanically so that an increase in the demand for shaft work will open all of them. Extraction pressure regulation is then achieved by moving upstream and downstream valves in opposite directions, to avoid changing shaft work.

The rangeability of these turbines is fairly broad. Extraction flow can be reduced to zero or increased to carry almost the full throttle flow. Only a small flow of cooling steam—perhaps 3% of the total—must flow through all stages. This is maintained by a low limit on the extraction throttle valves.

It is possible to lock the extraction throttle valves in a wide-open position, and either float the turbine on the steam headers or insert control valves in the extraction lines. Then load control would be entirely left to the inlet throttle valve, and header pressure regulation would be independent of the

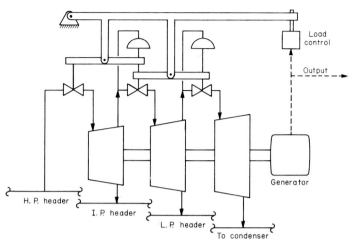

Fig. 3.1. Extraction turbines are essentially split into sections, each with its own throttle valve. The valves are mechanically linked for combined load control and pressure regulation.

turbine. Alternately the throttle valves could be operated independently of one another, so that electrical load control could be assigned to the last valve, for example. This type of operation is discussed under Section A.4.

3. Controlling Shaft Power

The shaft power delivered by a turbine is the product of speed and torque. The type of load being driven principally determines how these components vary. For example, an alternating current generator is fundamentally a constant-speed device. If operated in isolation from all other generators, its speed and hence the frequency of its generated power can vary. But each generator represents a tiny fraction of the total mechanical inertia on the electrical grid to which it is connected. If no shaft power is applied to an ac generator connected to the line, it acts as a motor. Therefore, the ac generator always operates at the same speed, as long as it is connected to the line, and its power output varies directly with the applied torque. Opening the throttle valve increases kilowatt output without affecting the speed. However, prior to connecting to the line, the speed of the generator must first be synchronized to line frequency.

Constant-torque loads are typified by hoists and conveyors which transport a uniform mass of material over a fixed distance. Production rate is then determined by speed, which is regulated by the steam throttle. The power required to lift a load is the product of its weight and velocity. Therefore, in a constant-torque system, power is directly proportional to speed.

Fluid propulsion tends to be neither constant-torque nor constant-speed. If a fluid were moved against a dominant static head, the power requirement would approach the product of pressure difference and flow. However, resistance to flow develops a dynamic head loss which is proportional to flow in the viscous regime and to flow squared in the turbulent regime.

The flow developed in a pipe is directly proportional to the speed of the prime mover, and the torque is proportional to the head it is working against. If the dynamic head varies with flow squared, the power required from the prime mover varies with flow cubed. Consequently, power also tends to vary with speed cubed. Naturally any static head affects the relationship as well. Load curves are developed for certain compressor applications in Chapter 4.

In any case, control of flow or pressure in a pipe is normally achieved by manipulating the speed of the turbine driving the pump or compressor. The turbine governor then responds by adjusting the throttle until the necessary speed is attained.

4.   Electrical Load Controls

In the central-station power plant, each turbogenerator must be controlled to meet the electrical demands placed on it by the dispatcher. In an industrial power plant, a specific demand for electrical power should *not* be placed on the generating system, unless it is independent of the electrical grid. If connected to the grid, such that power is being either purchased or sold, electricity generated within the industrial facility should be optimized on an economic basis.

The first economic choice is to minimize the steam being sent to condensers, as this represents heat loss. Then condensing turbines should not be operated except under abnormal conditions. Extraction turbines should operate with their last throttle valve set in its minimum position. That valve should not open until the next upstream valve is closed, in which case there is an excess of low-pressure steam to be used. It is preferable to use that steam to generate power than to vent or condense it without doing work. However, low-pressure is more valuable for heating than for work. Figure 1.7 indicates that 50 psia steam contains about 350 Btu/lb of available work, only 70% of which, or 245 Btu/lb, is recoverable in the turbine. If used for heat, however, 924 Btu/lb can be recovered by condensing it at its saturation temperature of 281°F. The recoverable work only represents 26.5% of the recoverable heat, which is below the efficiency of central-station power plants. In other words, it is more economical to purchase power than to generate it with low-pressure steam, if that steam can be used for heating— even to heat feedwater. However, if there is an excess of low-pressure steam, as indicated by a rising header pressure, it should be sent to the turbine.

Ideally, all heating steam (except that which must be at the highest available pressure) should be taken from turbine exhausts. Therefore, both intermediate and low-pressure headers should be supplied from turbines, either a double-extraction turbine or individual backpressure turbines. If individual backpressure turbines are used for electrical generation, their throttles should be manipulated to control the pressures in the headers into which they discharge. In most cases, other turbines, which are under control from the process they drive, will operate in parallel with those generating electricity. Then as either the work load or heating load shifts, the header pressure controllers will force the generating turbines to adjust accordingly. The generated power must then be expected to swing frequently and extensively to keep the header system in balance. The electrical grid then acts as a reservoir that accepts the normal fluctuations in the heat and work load of the plant.

There are obvious constraints to this mode of operation. Throttle valves could be forced closed or fully open, in which case pressure control would be lost. Then pressure-reducing valves may be opened as described later in the chapter. On the electrical side, the capacity of an individual generator could be exceeded, so that limits have to be applied.

A very important consideration is the "demand charge" assessed on purchased power by the utility. In addition to rate charges, in which energy is priced on a kilowatt-hour basis, a demand charge is assessed against the monthly peak kilowatt flow, as averaged over short (e.g., 15-min) demand intervals. The purposes of the demand charge are to provide incentive for users to level their loads, and to pay for the capital costs of the power plant facilities needed to meet peak demands (2).

Generating power within an industrial plant will not only reduce the purchased kilowatt hours but it can also shave the peaks from the kilowatt usage. If the turbine-generators are always under header-pressure control, plant steam usage will be optimized, but without regard to the demand for purchased power. The "spinning reserve," represented by a generator synchronized with the line but producing below its capacity, can be used to shave these peaks. The reduction in demand charge achieved in this way can usually offset a temporary shift away from an optimum steam balance.

Electrical generation with an extraction turbine can be increased simply by opening the throttle to the last stages. Steam diverted there from the low-pressure header would cause a reduction in header pressure which would open the other throttle valves and further increase generation. Or the mechanical linkage could open all valves directly. In either case, header pressures would still be regulated.

If the electrical generation is accomplished by a backpressure turbine, increasing its generation will sacrifice header-pressure control. In this case, control of pressure must be transferred to the feedwater heater, as shown in Fig. 3.2. As long as the use of purchased power lies below the demand limit, the signal from the kilowatt controller to the high selector will be low, and therefore not selected. But when generation must increase to limit purchased power, the throttle valve is directed to open further. Then the difference between the throttle valve position and pressure controller output is imposed on the feedwater heater valve. The pressure controller then effectively operates that valve. The pressure controller should be adjusted for its normal function of operating the throttle valve. Then the gain G of the subtracting device should be adjusted for control-loop stability when the heater valve is being manipulated. This technique of automatically changing control-loop assignments is described in Ref. (3).

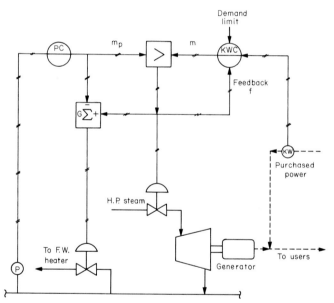

Fig. 3.2. When electrical generation is required to increase, header-pressure control must be transferred to the feedwater heater.

Under ideal conditions, the turbine is under pressure control, generating enough power to keep the purchased power well below demand. The kilowatt controller is then not in command, but needs to be ready to take command should purchased power rise to meet the demand limit. This conditioning is performed through external feedback. The kilowatt controller typically has proportional and integral action such that its output m responds to deviation e between purchased power and set point:

$$m = \frac{100}{P}\left(e + \frac{1}{I}\int e\,dt\right) \tag{3.1}$$

where P is the percent proportional band and I is the integral time constant. (For background information on controller modes, the reader is urged to consult Ref. 4.) The integral function in some controllers is provided by positive feedback of the output through a first-order lag of time constant I, as shown in Fig. 3.3. In the steady state, the input and output of the lag will be equal, in which case e must of necessity be zero. If e is other than zero, the output is continuously being changed by the positive feedback.

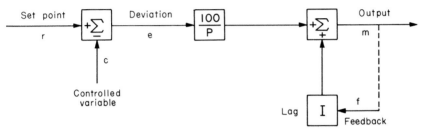

Fig. 3.3. Integral action in a controller may be achieved by positive feedback, allowing it to be stopped by opening the feedback loop.

If the positive feedback loop is opened at the broken line, and another signal f inserted in place of m, a steady state will be reached when

$$m = f + (100/P)e \tag{3.2}$$

In order for the controller's integral mode to function, f must equal m, but if it is necessary to stop integral action, f should not equal m. In the system shown in Fig. 3.2, feedback f is equal to m only when m is selected to operate the throttle valve. If the pressure controller needs more steam to maintain header pressure than the kilowatt controller does to meet electrical demand, the output of the pressure controller m_p will be selected for throttling. Then m_p is fed back to the kilowatt controller, bringing its integrating action to a halt. Note that $m = m_p$ at the point of selection, in which case Eq. (3.2) indicates that the kilowatt deviation will be zero.

The pressure controller does not require conditioning by feedback. If it is not manipulating the throttle valve, it is manipulating the heater valve— therefore, it is always in control.

5.   Controlling Multiple Turbines

Manipulating more than one turbine to control electrical generation
complicates the system considerably. The plant operator must be able to
balance the load between generators as he sees fit. The bias stations ($\pm$)
shown in Fig. 3.4 are provided for this purpose; they also should be equipped
with adjustable limits and a manual mode of operation to permit base-
loading of each turbine.

To prevent the balancing adjustments or imposition of limits from
upsetting generated power, the weighted average position of the throttle
valves is fed back as measurement to a valve-position controller. Because
this controller only encloses instruments in its loop, it can be adjusted for
very rapid response—an integral time constant of less than 1 sec is possible.

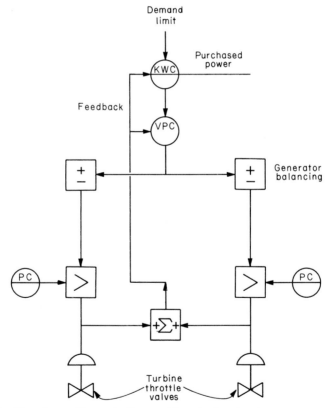

Fig. 3.4. The valve-position controller repositions the throttle valves very quickly following
a change in the output of the kilowatt controller, either pressure controller, or a balancing
adjustment.

Then any change affecting a valve is quickly acted upon to adjust the remaining valve(s) to maintain the previous average. If the installed characteristics of the valves are reasonably linear, or if they are operating in the same portion of their travel, little upset to generated power should result from such an adjustment. Additionally, failure of any valve to respond to control action because it is in manual or under pressure control affects only the gain of the very fast valve-position loop. The response of the kilowatt control loop is essentially uniform for all conditions.

Should *no* valves be under kilowatt control because of pressure control override or other reasons, the valve-position controller will sustain a steady-state deviation. By using the valve-position signal as feedback to the kilowatt controller, its integration is stopped by a valve-position deviation. The KWC is therefore always conditioned properly; the VPC is not, but its extremely rapid response effectively eliminates the need for conditioning.

6. Choked-Flow Control

When the velocity of the vapor leaving a turbine exhaust exceeds the tip velocity of the blades, further increases in flow produce comparatively little increase in shaft power. This condition, known as "choked flow," represents the practical upper limit of exhaust velocity.

The two factors affecting exhaust velocity are the mass flow of steam from the last stage, and its absolute pressure. Mass flow is determined by load and is not ordinarily controllable, although exhaust pressure is controllable. The pressure in the header into which a noncondensing turbine discharges could be increased in proportion to mass steam flow to keep the velocity from increasing above choked flow. With a condensing turbine, condenser pressure could be raised by reducing the flow of cooling water to accommodate an increased mass flow.

Specifically for a turbogenerator, which operates at constant speed, mass flow is linear with absolute inlet pressure. Then exhaust velocity will be proportional to the ratio of absolute inlet pressure to absolute exhaust pressure. A ratio controller can then be used to keep exhaust pressure in a certain ratio to absolute inlet pressure. An exhaust pressure higher than this ratio will reduce turbine power, whereas an exhaust pressure lower than this ratio will not result in any significant increase.

B. STEAM HEADER CONTROLS

To this point, the effects of backpressure turbines and header pressure on each other have been explored. Pressure control using these turbines has also been described. There are many other considerations, however, principally stemming from the demands for heat at various pressure levels in the plant. Only after all of these factors have been taken into account can an effective header-management system be devised.

1. Regulating Header Pressure

Steam headers are actually low-capacity reservoirs of energy distributed through the plant. Flows are continually entering and leaving through various paths and at changing rates. A balance of inflows and outflows is rarely reached without being forced by a pressure controller. When demands exceed receipts, the controller must deliver more steam from a higher-pressure source, either through a turbine or let-down valve. (The exception is the highest-pressure header, which is supplied directly from the boiler and whose control is subject to heat-input manipulation.) When receipts exceed demand, some steam must be spilled to a feedwater heater, condenser, or vent.

If control is not enforced, the pressure will rise or fall in an attempt to regulate itself. During a heavy demand, falling pressure will reduce flows to all users. As long as the controllers on those users can react by opening their valves more, they can still be satisfied and pressure will continue to fall. Only when one or more valves are fully open or otherwise limited will flow begin to decrease and pressure begin to stabilize. The converse is true of an oversupply and subsequent rising pressure. Needless to say, if a balance is not reached, at least some of the processes using the steam will be adversely affected.

Regulation of pressure with simple reducing and relief valves is effective but also wasteful. To conserve energy, priorities must be assigned to sources and users of steam in order of their cost. Because work has already been extracted from turbine-exhaust steam, it should be used to the fullest, as described in Fig. 3.2. However, high and low limits of steam availability may be encountered in that source, and the control system must be capable of reacting to those limits. If more steam is required than the turbine(s) can supply, only then should steam be let down through a valve from a higher-pressure source.

Priorities also exist for rising pressure. If the turbine throttle is closed or limited by the kilowatt controller, steam must be released, first to heat feedwater, then to a condenser, and vent in that order.

Figure 3.5 shows a method for controlling header pressure by manipulating a noncondensing turbine within limits. This system differs from that shown in Fig. 3.2, primarily in the addition of the let-down valve and its sequencing with the heater valve.

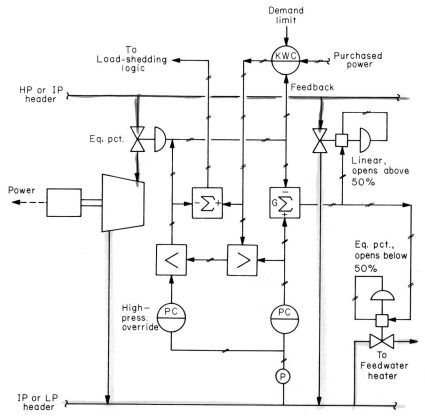

Fig. 3.5. Header pressure should be controlled by manipulating a noncondensing turbine-generator; bypassing the turbine with a let-down valve or using steam to heat feedwater should only be resorted to when the turbine is limited.

Since let-down valves pass steam between headers which are maintained at constant pressure, they should have a linear characteristic. Turbine throttle valves, on the other hand, face constant upstream pressure, but downstream pressure varies with flow. Therefore, they require the equal-percentage characteristic. The same is essentially true of the heater valve. To increase heat transfer through the heater surface, its temperature difference must rise proportional to flow. Thus the pressure at which the steam condenses in the heater will change with flow.

The let-down and heater valves are operated in split range through the action of their positioners. Both valves are closed at a 50% control signal, with the let-down valve opening above 50% and the heater valve opening below. The subtractor is then calibrated to produce a 50% output when its two input signals are equal.

Relief of excess steam to any other sink except a feedwater heater merits careful consideration. If it is discharged to a lower-pressure header it will surely become an excess there. The problem is thereby not solved but merely passed on. The natural consequence of relieving steam from a higher- to lower-pressure header is a cascading of flows to the lowest-pressure header and on to the condenser or vent. In effect, the excess steam at the point of origin is being sent to the condenser or vented.

Rather than waste steam to condenser or vent, a high-pressure override should limit power generation by the turbine exhausting steam into that header. Then the kilowatt controller would fail to keep purchased power below demand, and load-shedding must be initiated. An independent pressure controller having a higher set point is used for the override function as shown in Fig. 3.5. It would allow header pressure to float somewhat above the normal set point before initiating an override. This will increase the capacity of the header to absorb momentary upsets, and also force more steam into the feedwater heater.

When the override controller assumes manipulation of the turbine throttle, a difference will appear between the signals connected from the low selector to the subtractor. Thus the output of the subtractor represents deficient generating capacity, and can be used as a load-shedding signal. If the override controller has an adjustable high limit, that limit can be used to set the maximum steam flow to the turbine. Any departure of the kilowatt-controller output above that limit will also develop a proportional signal for load-shedding.

Should the override controller be incapable of limiting pressure rise, having closed the turbine throttle to its low limit, the situation has become unmanageable. Then an alarm should be sounded and excess steam relieved to a vent or condenser.

2.   Desuperheaters

Steam that is let down from a higher pressure becomes superheated at the lower pressure. Superheat is desirable for work-producing equipment but not for heaters. Therefore, desuperheating should be applied to any let-down stream that will be used for heating.

Desuperheating is accomplished by mixing the superheated steam with condensate to approach a more saturated condition. Mixing has been cited as an irreversible process which should be avoided, but pressure reduction through throttling is also. Irreversibility in the desuperheater can be minimized by using condensate which most nearly matches the saturation temperature of the steam.

To produce 1 lb of saturated steam at enthalpy H requires x_1 pounds of superheated steam at H_1 and $(1 - x_1)$ pounds of water at H_2:

$$H = H_1 x_1 + H_2(1 - x_1) \tag{3.3}$$

Rearranging,

$$x_1 = (H - H_2)/(H_1 - H_2) \tag{3.4}$$

The higher H_2 is, the smaller x_1 will be. Therefore, the least flow of super-heated steam will be required when H_2 corresponds to saturated liquid at the operating pressure.

An entropy balance bears this out, as would be expected. Table 3.2 lists the required flow of steam at 600 psia and 600°F needed to produce 1 lb of saturated steam at 150 psia. Remember that the steam is being let down to provide the needed flow of saturated steam at the lower pressure—not to relieve excess high-pressure steam. It should be noted that the entropy of mixing is far less than the entropy gained by letting down the steam. The contribution of throttling 0.9 lb of steam from the stated initial to final conditions is 0.1164 out of the 0.1390 Btu/°F given in the last column.

Table 3.2

Requirements of 600-psia, 600-°F Steam to Make One Pound of 150-psia Saturated Steam

Water temperature (°F)	Water enthalpy (Btu/lb)	Steam req'd (lb)	Entropy increase (Btu/lb-°F)
358[a]	330.5	0.900	0.1390
212	180.1	0.914	0.1427
60	28.0	0.924	0.1493

[a] Saturated at 150 psia.

Figure 3.6 shows a desuperheating system using condensate from a high-pressure heater; heated boiler feedwater could be used if no high-pressure condensate is available. The mixing is best achieved in a three-way steam-conditioning valve such as that offered by Yarway Corporation (5). As the pressure controller opens the steam inlet port, the water inlet port opens proportionally, providing a regulated amount of desuperheating. An additional valve on the water supply line is manipulated to reach the desired final temperature.

Lack of availability of condensate, as indicated by a low level in the condensate pot, would cause a reduction in allocation for desuperheating. The level controller output is amplified by a factor G and passed to a low

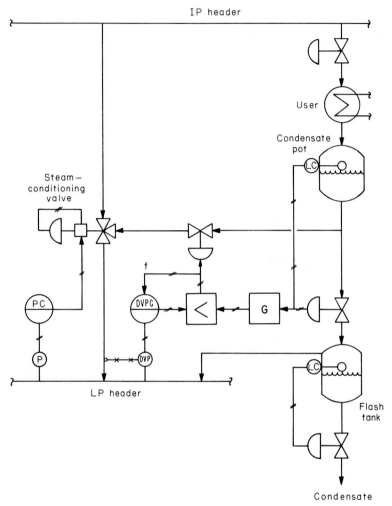

Fig. 3.6. Let-down steam should be desuperheated with the hottest available condensate; differential vapor-pressure control will maintain superheat under conditions of variable pressure.

selector to be compared with the output of the superheat controller. The value of G is set high enough to avoid interfering with superheat control except when the level is very low.

Superheat is measured in this system by a differential vapor-pressure device (6). This device is essentially a differential pressure transmitter, one side of which is a filled thermal system. The filling fluid—in this case water— exerts its vapor pressure against the measured header pressure on the other side of the cell. If the steam is saturated or wet, pressures on the two sides will be equal; superheat is indicated by a vapor pressure higher than measured pressure. Superheat is measured in terms of pressure rather than temperature, but the two are related linearly for any given pressure. Figure 3.7 plots that relationship for three pressures.

The reasons for controlling differential vapor pressure instead of temperature are its superior sensitivity (as indicated by Fig. 3.7), and its ability to accommodate moderate variations in pressure. Being able to vary header pressures is one of the more important aspects in achieving steam economies.

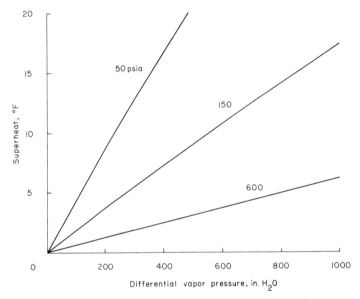

Fig. 3.7. Differential vapor pressure is linear with superheat at any given pressure.

3. Header-Pressure Optimization

The header pressures in most plants were probably assigned arbitrarily during the early stages of design and never changed. They are among the "untouchables" in a plant, to be accepted rather than adjusted or even questioned. However, plants rarely if ever operate at the conditions that were used by the designers who assigned those pressures. At any point in time, that assigned pressure might be too low for one or more users, or far higher than necessary for all users.

Production rate has a profound influence over this relationship. When production is at its highest, the assigned pressure may not be high enough to satisfy all the users, as indicated by wide-open control valves. Yet when production is low, valves are throttled and perhaps could accept a much lower pressure.

Recognizing that throttling is an irreversible process, it would seem desirable to force control valves farther open, yet without losing control of their individual variables. This can be done by reducing header pressure as production rates cause valves to close. In order to satisfy all users, however, the pressure must be held at a point that will keep the most open valve just short of fully open.

The system shown in Fig. 3.8 accomplishes this. The positions of all the control valves supplied by the header are compared in a high selector. The position of the most open valve is then maintained at 90 or 95% open by a controller which adjusts the set point of the header-pressure controller, between specified limits. Rapid fluctuations in pressure are not allowed, as they will upset all the users. Therefore, the valve-position controller must do its job slowly, having little or no proportional action, and an integral

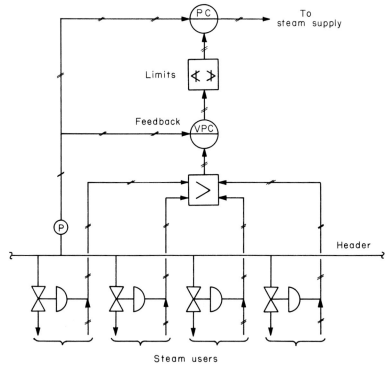

Fig. 3.8.   Header pressure should be as low as possible, consistent with satisfying the most-open valve.

time of several minutes. Because of its slow response, it cannot hold the selected valve position tightly at the set point, but only keep its average position there. Then short-term upsets will be quickly countered by the action of the pressure controller, while long-term variations in load will cause header pressure to rise or fall slowly. Using the measured pressure as feedback to the valve-position controller will properly condition it whenever limits are encountered, or when pressure is controlled at a locally set value.

This system will keep the header pressure as low as possible consistent with steady-state loads. The virtues of minimum-pressure operation were pointed out in conjunction with Table 3.1. Reducing turbine backpressure from 150 to 120 psia increased the work recovery by 27.5%, without incurring any other debit. The heaters using this steam would notice its enthalpy reduced by less than 20 Btu/lb in 877 (the latent heat of 120 psia steam)— hardly a significant factor. In fact the steam they use would become less superheated at the lower pressure and reduced throttling—so all users would benefit.

A turbine having a variable-pressure supply would be unaffected as long as its throttle valve is less than fully open. This was also discussed earlier. Therefore, *all* header pressures should be at their minimum level, consistent with user needs as achieved by the system of Fig. 3.8.

4.   Accommodating Variable Pressure

Some users can accept variable steam pressure if the variation is slow enough. Simple heat exchangers which are under temperature control are capable of correcting for an upset in supply pressure well within a minute. The valve-position controller of Fig. 3.8 ought to be adjusted with an integral time of several minutes—perhaps even an hour. Then it would be incapable of promoting a change in pressure that would be rapid enough to upset the variable under control.

Similarly, liquid level is quite often controlled by manipulating the heat input to vaporizers and reboilers. This variable responds even more rapidly than temperature to changes in heat input, and therefore would not tend to be affected by gradually varying pressure.

The steam to many processes is under flow control. Then a change in steam pressure that is large enough to cause a flow change will be acted upon by the flow controller to restore flow to the set point. However, flowmeter calibration is also affected by pressure.

Orifice meters are the most common elements for measuring steam flow. The differential pressure h developed across an orifice is related to mass flow M as

$$M = k\sqrt{h/v} \tag{3.5}$$

where k is a scaling constant and v is the specific volume of the steam. If steam were a perfect gas, its specific volume would be related to absolute temperature T and pressure p as

$$v = RT/wp \qquad (3.6)$$

where R is the universal gas constant in appropriate units, and w is the molecular weight of water, 18. But because steam is far from ideal, its specific volume can only be approximated over limited ranges of temperature and pressure as

$$v_c = (a + bT)/(c + p) \qquad (3.7)$$

Coefficients a, b, and c can be found to match three points of v versus T and p, bracketing the expected operating range. If the thus-calculated values of v_c do not satisfactorily agree with tabular values of v in other important areas, then a, b, and c should be readjusted for a better fit.

For saturated steam, a plot of specific weight $1/v$ against pressure forms a nearly perfect straight line at low pressures, as seen in Fig. 3.9. In this case, the mass-flow equation reduces to

$$M = k\sqrt{h(ap + b)} \qquad (3.8)$$

where a and b are selected for the best fit.

In many, if not most, processes, however, *heat* flow is more important than *mass* flow. Then variations in enthalpy as well as specific volume should also be taken into account. The heat flow through a steam flowmeter can be expressed as

$$Q = \Delta H M = \Delta H k \sqrt{h/v} \qquad (3.9)$$

where ΔH is the heat given up in condensing the steam at the metered pressure. If the steam actually condenses at a lower pressure due to the pressure loss across the control valve, or if the condensate is subcooled, more than ΔH will be given up. Nonetheless, Eq. (3.9) will give a more accurate result than applying no compensation at all.

The group $\Delta H^2/v$ was factored from Eq. (3.9) and plotted against pressure for saturated steam in Fig. 3.9. Although its plot is not as linear as that of $1/v$, it can still be linearized accurately over limited operating ranges. The form of Eq. (3.8) may be applied, selecting coefficients a and b to fit the operating region:

$$Q = k\sqrt{h(ap + b)} \qquad (3.10)$$

If the steam is superheated, increasing temperature (at constant pressure) raises both ΔH and v. As it happens, the simultaneous variations in the two terms bring about nearly complete cancellation. Therefore, Eq. (3.10) can be used for both saturated *and* superheated steam, again with a and b selected

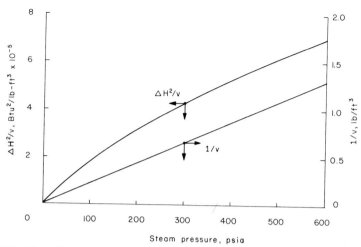

Steam pressure, psia

Fig. 3.9.   The orifice correction factor for mass flow is linear with saturated steam pressure, while the correction factor for heat flow exhibits some curvature at lower pressures.

for the best fit in the operating region. Figure 3.10 describes how pressure-compensated steam flow would be applied to an evaporator to control product composition.

Certain processes having extensive capacity can be used as scavengers of low-pressure steam. One that is commonly employed in the food-processing industry is the multieffect evaporator. Insufficient steam may be available

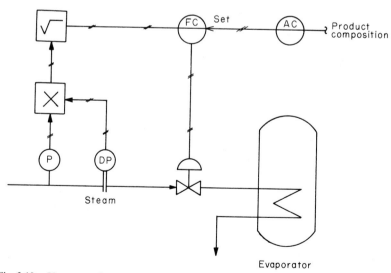

Evaporator

Fig. 3.10   Users can be protected against supply pressure variations by manipulating pressure-compensated steam flow instead of valve position.

at any point to produce evaporation at the desired rate. However, the control system can be designed to reduce feed rate in proportion to steam availability.

In essence, the steam flow to the evaporator would be set proportional to the liquid level in the feed tank, and feed rate set proportional to steam flow. Then an interruption in steam flow would cause a commensurate reduction in feed rate, thereby maintaining control over composition, but losing control over feed-tank level. However, feed-tank level does not really require control—it only serves as an indication of inventory and a request for average production. When steam flow is restored, production resumes at a rate proportional to tank level. If the tank is capable of holding an hour's inventory of feedstock or more, steam shortages lasting several minutes may be accommodated with ease. The only requirement is that the average availability of steam must at least match the average feed rate. This system is outlined in detail in Chapter 6.

C.   BOILER CONTROLS

The boiler is required to respond to the changing demands of a multiplicity of steam users. How well it can meet these demands depends on the responsiveness of its controls and the rapidity with which these demands vary. Heat capacity and water storage between the fuel and the steam user are the ultimate limiting factors, assuming that the control system performs its function.

However, response to demand is of itself insufficient—the manipulated flows of fuel, air, and water must be balanced to satisfy the critical levels, pressures, and temperatures in the system. The controls described here are devised to maximize response while maintaining these balances. Feedforward control is used extensively in boilers to achieve the requisite combination of balance and response.

Boilers are available in two basic designs: fire-tube and water-tube. Fire-tube boilers are generally limited in size to about 25,000 lb/hr, and 250 psig, saturated steam. While they are noted for their ability to respond to changing demands, their size and pressure limitations preclude their use in large industrial facilities. Due to thermodynamic considerations, boilers should produce steam at high pressure and temperature, to realize a maximum work efficiency. These conditions are only achievable with water-tube boilers— hence they are given prime consideration.

1.   Drum-Level Control

The steam drum functions to disengage steam from the two-phase mixture rising from the evaporating section of the boiler. The liquid phase is returned through a downcomer to absorb more heat, completing the natural-circulation pattern as shown in Fig. 3.11. At pressures above 2000 psig, the density difference between vapor and liquid decreases to the point where natural circulation may not provide the velocities needed for efficient heat transfer. Then forced circulation is required. Some lower-pressure units that are expected to respond rapidly to load changes such as combined-cycle units (7) also used forced circulation. But whether the circulation is natural or forced, the drum-level control problem is the same.

Aside from thermal considerations, liquid level is relatively easy to control by manipulating the inflow to the vessel. However, the boiling of the two-phase mixture into the drum causes spontaneous fluctuations in measured level which are too rapid to be affected by control action. These disturbances are referred to by the control engineer as noise.

Hydraulic resonance is also common to vessels with an open surface of liquid. The liquid can oscillate from side to side like a glass of water that has

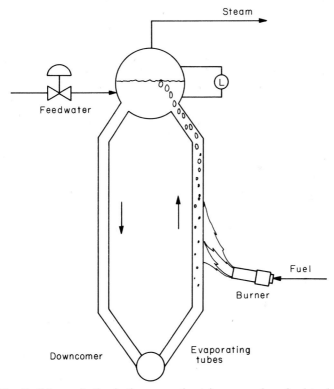

Fig. 3.11. Partial vaporization in the evaporating tubes causes drum level to shrink when feedwater flow increases and when pressure rises.

been disturbed, and it can also cycle vertically in the gage glass as is seen in a large coffee urn. This resonance is governed by the same principal as a pendulum (4, pp. 71–73), having a natural period

$$\tau_o = 2\pi\sqrt{L/2g}$$

where L is the distance in feet between the bounded surfaces, e.g., the two sides of the vessel, or the level in the vessel and the level in the gage glass. Constant g is the acceleration of gravity at 32.2 ft/sec^2 and τ_o is expressed in seconds. As the dimensions within the drum lie between 2 and 20 ft, the periods observed are between 1 and 3.5 sec. While the resonance is damped by resistance to flow, the continuous disturbances introduced by the rising stream from the evaporating tubes are reinforced at the natural period.

Feedwater is always colder than the saturated water in the drum. Some steam is then necessarily condensed when contacted by the feedwater. As a consequence, a sudden increase in feedwater flow tends to collapse some bubbles in the drum and temporarily reduce their formation in the evaporating tubes. Then, although the mass of liquid in the system has increased, the apparent liquid level in the drum falls. Equilibrium is restored within seconds and the level will begin to rise. Nonetheless the *initial* reaction to a change in feedwater flow tends to be in the wrong direction, as shown in Fig. 3.12. This property, called "inverse response," causes an effective delay in control action, making control more difficult. Liquid level in a vessel lacking these thermal characteristics can typically be controlled with a proportional band of 10% or less. By contrast, the drum-level controller needs a proportional band more in the neighborhood of 100% to maintain stability. Integral action is then necessary, whereas it can usually be avoided when very narrow proportional band settings can be used.

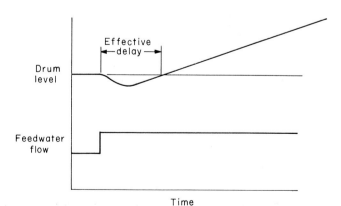

Fig. 3.12.   The inverse response of drum level to feedwater flow makes its control difficult.

Rapidly varying demands for steam also create problems in drum-level control. Naturally, an increase in steam generation will cause drum level to fall until it is matched by an equal increase in feedwater flow. Steam generation can be increased in two ways: by increasing the rate of firing, and by allowing boiler pressure to fall. Both of these mechanisms are employed in meeting increasing steam demands.

When a steam user opens its control valve, steam flow from the boiler to the user immediately increases. Heat transfer through the boiler cannot increase as quickly, even if fuel flow were augmented directly upon notification of the new demand. Consequently, steam pressure will begin to fall. The falling pressure causes some of the saturated water in the drum and evaporating tubes to boil, which temporarily satisfies the new demand. Eventually the increase in heat input will raise pressure to its original value, and equilibrium will be restored.

However, during the time that pressure is falling, increased evaporation raises the percent vapor in the evaporating tubes, physically lifting more water into the drum. The falling pressure thereby causes a "swell" in drum level—a false indication of the load change in that it develops in the wrong direction. Acting on the swell, the level controller would begin to decrease feedwater flow, when it should actually increase. The reverse happens on a falling demand—pressure rises, causing drum level to "shrink."

Feedforward control is used to maintain a steam–water balance, reducing the influence of shrink–swell and inverse-response phenomena. The system shown in Fig. 3.13 causes feedwater flow to match steam flow in absence of action by the level controller. The two flowmeters have identical ranges, and their signals are subtracted. If the two flow rates are identical, the

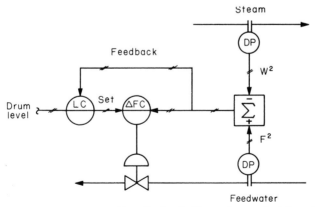

Fig. 3.13. The system above provides effective feedforward control of drum level with a minimum complement of instruments.

subtractor sends a 50% signal to the flow-difference controller. An increase in steam flow will call for an equal increase in feedwater flow, to return the difference signal to 50%.

Errors in the flowmeters, and the withdrawal of perhaps 2.5% water as "blowdown" (which is not converted to steam) will prevent the two flow signals from being identical. Any error in the steam–water balance will cause a falling or rising level. Therefore, the level controller must readjust the set point of the flow-difference controller to strike a steady-state balance. Since its role is reduced from manipulating feedwater flow across its entire range to adjusting only for flowmeter errors and variations in blowdown, deviations in level from its set point will be minimized. Then controller mode settings are not as critical, and incorrect actions caused by shrink, swell, and inverse response are reduced.

The flowmeters used for steam and feedwater are usually orifices or nozzles, which produce differential pressures varying with flow squared. The system shown in Fig. 3.13 does not use square-root extractors, because of an interesting dynamic-response characteristic often observed. The author discovered that the period of oscillation and dynamic gain of a two-capacity level process varies directly with flow (8). The gain of the feedwater control loop without square-root extraction seems to compensate correctly for the process gain change. The level controller essentially adjust the flow-difference set point in increments dh, where h is the measured differential pressure. In turn, feedwater flow responds in increments dF. Because

$$F = k\sqrt{h} \tag{3.11}$$

then

$$dF/dh = k/2\sqrt{h} = k^2/2F \tag{3.12}$$

Therefore, the gain variation of the feedwater-flow loop compensates that of the process.

Figure 3.13 also shows external feedback from the flow-difference measurement applied to the level controller. This will precondition the level controller during startup or other times when feedwater is controlled manually or otherwise limited. For extremely low loads and during startup, a second level controller is often used, manipulating a small feedwater valve directly. Figure 3.14 shows how the low-range controller and valve can be integrated into the high-range system.

When feedwater flow is low, the switches are in their left-hand position, connecting the low-range controller to the small valve. At the same time, the flow-difference controller is disconnected and the high-range level controller is preconditioned to its normal 50% output. When flow reaches the normal range, the switch is reversed and normal-range control begins.

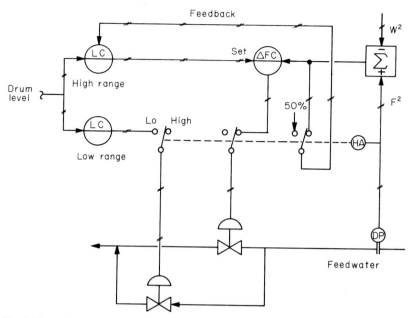

Fig. 3.14. When the flow of feedwater reaches the setting of the high-alarm, control is automatically transferred to the preconditioned high-range system.

2. Boiler-Pressure Controls

Just as the drum level is an indication of steam–water balance, drum pressure is determined by the balance between heat inflow and outflow. Consider opening a steam valve without a concurrent change in firing rate. Steam flow to the user will increase, but only temporarily; eventually the steam pressure will reach a new, lower level at which point steam flow will have returned to its former value.

Boiler pressure in an industrial facility can only be regulated by manipulating heat *input*, for heat *output* is subject to the combined demand of all the users. In order to bring about a change in pressure, the changing fuel flow must first affect the distribution of flue gas, and the temperature of the metal and refractory materials in the boiler. This involves a delay of several seconds to a minute, depending on the boiler and its load level. Feedforward control can improve pressure control by adjusting fuel as soon as a load change is detected, instead of waiting for pressure to change first.

The actual steam flow at any point in time is not necessarily a true indication of demand, however. If a demand increase takes place when steam pressure is at set point, the desired steam flow is obtained only momentarily.

As pressure falls, steam flow will also fall if the user valve or valves remain in position. A subsequent increase in steam flow caused by increased firing likewise should not be interpreted as a load increase—this would create a positive feedback loop, capable of destabilizing the boiler.

The true load placed on a boiler is the combined openings of all valves drawing steam. But rather than determining the load from a multiplicity of valve-position signals, it can be inferred from steam flow and pressure. Consider a combination of valves and/or turbines accepting high-pressure steam and discharging into a pressure half as high. The flow of heat through those combined orifices of area $\bar{m}$ can be estimated as

$$Q = k\bar{m}p \tag{3.13}$$

Presumably the heat flow is measurable by an orifice or nozzle per Eq. (3.10). Then the combined valve area $\bar{m}$ for all users may be found:

$$\bar{m} = Q/kp = \sqrt{h(ap + b)/p} \tag{3.14}$$

The plant load can then be approximated as

$$\bar{m} \approx \sqrt{h/p} \tag{3.15}$$

which is probably close enough for control purposes.

Figure 3.15 describes a boiler-pressure control system using this type of feedforward model. Fuel flow is set proportional to the estimate of load $\bar{m}$. Dynamic compensation is applied in the form of a lead–lag function to help overcome the heat capacity in the boiler. This function is described in detail in Ref. (4, pp. 211–219) and need not be repeated here.

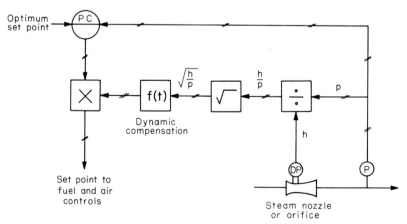

Fig. 3.15. Both steam flow and pressure are used to set the firing rate to the boiler.

The pressure controller adjusts the ratio of fuel flow to estimated load, to correct for inaccuracy of the model and variations in heat of combustion of the fuel. The multiplier also changes the gain of the pressure feedback loop proportional to load. This feature is valuable in that boilers seem more difficult to control at lower loads due to lower velocities.

Boiler pressure should not be held at some arbitrary value any more than header pressures. In fact, "sliding-pressure" operation has recently been incorporated into the design of some large central-station power plants. During low-load conditions, the pressure is reduced so that the turbine throttle can open wider, resulting in less loss of available work. In addition, the boiler feed pump then works against a lower discharge head, saving more energy. Pump power is worth saving in high-pressure boilers—as much as 3% of the gross work produced by a 2400 psig boiler is used to pump feedwater.

The mechanism for determining the optimum boiler pressure is the same as used for header pressures (see Fig. 3.8). As long as the turbine throttle valve is less than fully open, a reduction in supply pressure has no effect on the work it produces. But other than reducing feedwater pump power, the advantage of minimum boiler pressure lies in utilization of the steam by turbines, etc., as opposed to improving boiler efficiency, which it does not. If a maximum of electricity should be generated, then the pressure should be high. Only during conditions of low work loads should the pressure be lowered, again consistent with positioning of the user valves.

3. Superheat and Reheat

Superheat and reheat increase the available work in the steam above what is attainable at saturation. In fact, the highest work efficiencies can only be reached by maintaining the highest possible steam temperatures that the metals are capable of carrying. Central-station steam temperatures are limited to about 1050°F, while those in industrial units may be considerably less. If these temperatures can be controlled with extreme precision, they can be pushed closer to the allowable limits. For example, excursions of ±50°F would mean that the set point would have to be held at 1000°F; if the variation could be reduced to ±10°F by more effective control, the set point could be elevated to 1040°F. This is enough to increase the enthalpy of 2400 psia steam by 25 Btu/lb and its available work by 17 Btu/lb.

As steam is expanded through a high-pressure turbine, it loses superheat. In an effort to avoid excessive condensation in lower-pressure turbines receiving that exhaust, it is reheated to the same temperature as it was originally superheated. At this point in the cycle its pressure is much lower, however, typically 600 psig for a 2400-psig boiler.

Superheat and reheat can be controlled by two different mechanisms. Feedwater may be sprayed into the steam in an "attemperator" to reduce its superheat. Since this process is irreversible, it wastes available work, and is therefore to be minimized. Attemperation is held at a low value in actual practice, being perhaps 2% of feedwater flow at unfavorable loads, to zero under more favorable conditions.

The fuel–air ratio in a boiler does not change greatly across the entire load range; as a consequence, flame temperature does not change significantly with load either. But the flow of fuel and air increase with load, causing the hottest gases to propagate farther through the boiler. If no corrective measures are taken, the temperature of the flue gas encountering the superheat and reheat sections at low load are too low to transfer any heat. Some boilers have tilting burners which can be directed toward these sections at low load and progressively downward as the load increases. Large coal-fired boilers have recirculation fans that return flue gas back to the combustion zone to more equitably distribute gas temperatures. As the load increases, fuel and air flows increase, requiring less recirculation. Finally a load is reached where too much heat is directed to superheat and reheat sections, and a damper must be opened to bypass gas around them.

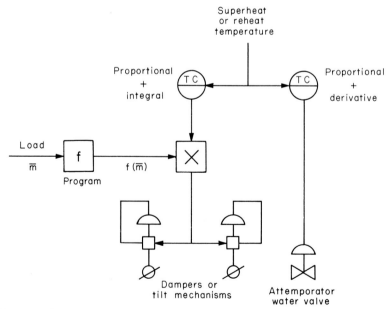

Fig. 3.16. Steam temperature control is separated into two components, with integral action applied to flue-gas manipulation and derivative applied to attemperation.

The required positions of burners or recirculating and bypass dampers as a function of load is well established for any given boiler design. Therefore, it is common practice to program their positions directly from load. Naturally, some readjustment is necessary to correct for inaccuracies in the program and changes in the characteristics of the boiler. This is accomplished by feedback control of temperature, using proportional and integral action. Being applied through a multiplier, the feedback loop gain varies directly with load, tending to cancel the inversely varying process gain (9).

Temperature control by attemperation is more responsive and can be used to supplement flue-gas manipulation. To minimize water usage, however, and to avoid conflict with flue-gas manipulation, proportional-plus-derivative control should be used for attemperation. The controller may be biased to deliver a nominal amount of feedwater at zero temperature deviation. The control system is described by Fig. 3.16.

D. MULTIPLE-BOILER PLANTS

Most industrial steam plants have been expanded incrementally through the years so that their present capacity resides in several boilers rather than one. This distribution of load-carrying capacity can be both a blessing and a curse. Its advantages include the possibility of "base-loading" some units at their most efficient conditions, while following demand with other faster-responding boilers. It is possible to maximize efficiency and schedule preventive maintenance when multiple units are available. However, the very flexibility it affords presents too many management decisions. Determining the optimum load distribution for five boilers is much more difficult than balancing two. Similarly the multiple-boiler control system must be capable of responding equally well regardless of the number of units in service, while accommodating changes in load distribution and availability. These are difficult problems, but they must be solved if the advantages of multiple-boiler installations are to be realized.

1. Waste-Heat Recovery

Scattered through a petroleum, chemical, or metallurgical complex may be a number of boilers which are used to recover waste heat from process streams. Some of these boilers may be fired in that the heat is available from the combustion of a feedstock or waste product. For example, smelting of sulfide ores is a combustion process whose energy is readily recovered by a waste-heat boiler. Similarly, carbon monoxide from the regeneration of catalyst in an oil refinery must be incinerated, and the resulting heat is usually recovered in a boiler. These combustion systems may require supplementary fuel for startup or to avoid flameout.

Quite often the sensible heat of products leaving a reaction system is recovered in a waste-heat boiler. While using this energy to preheat feed or to heat boiler feedwater is perhaps more common, the waste-heat boiler has the advantage of superior temperature regulation owing to its isothermal operation.

Regardless of the source of waste heat, recognize that its availability depends on the operating conditions of the process. Therefore, it is a slave to the process—the heat must be removed as it is generated in order to maintain control over the process. Waste heat may or may not be available when it is needed. In any case, the steam header system must accept it when it is available, while still being capable of functioning without it.

A particular difficulty arises with batch processes which produce copious amounts of waste heat—a copper reverberatory furnace is a good example. Before the process can be started, heat is required in the way of supplemental fuel to raise temperatures, and work to operate air compressors. If a compressor is steam-driven, it must be supplied by an auxiliary boiler. As the reaction proceeds, the waste-heat boiler begins to generate steam, and the auxiliary boiler can be cut back. Eventually more steam may be generated than needed to drive the compressors, and another use must be found for the excess. Finally the reaction approaches completion, with steam availability and need both decreasing. Usually a plant will have several of these batch processes operating in parallel. They should be scheduled to minimize the use of fuel required for the auxiliary boiler, and the waste of steam above what can be used to drive compressors and generate electricity.

Those waste-heat boilers which supply steam directly into the high-pressure header can upset that pressure as process conditions change. The measured steam flow from each should be subtracted from plant load to properly estimate the demand for fuel. Figure 3.17 shows how these signals would be incorporated into the pressure-control system described by Fig. 3.15. Note that the differential pressure signals must be linearized before addition or subtraction. In this system, plant load $\bar{m}$ is calculated as in (3.15), using the sum of the flow measurements. Then the heat flow demanded of the fired boiler is calculated by subtracting the measured heat flow of the waste-heat boilers from the plant load.

When steam from a waste-heat boiler is injected into a *header*, provision should be made to accept variations in its flow without upsetting header pressure. Accordingly, the flow of steam should be measured, pressure-compensated if header pressure is variable, and subtracted from the output of the header-pressure controller as shown in Fig. 3.18. The output of the subtractor, representing the demand for additional steam from a higher-pressure source, would be sent to the turbine control system. In the system shown in Fig. 3.5, for example, the subtractor would be inserted directly in the output of the header-pressure controller.

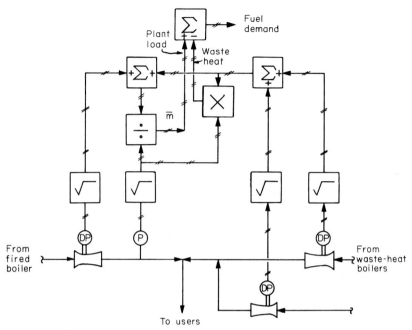

Fig. 3.17. The steam flow from waste-heat boilers must be subtracted from estimated plant load to calculate the demand for fuel.

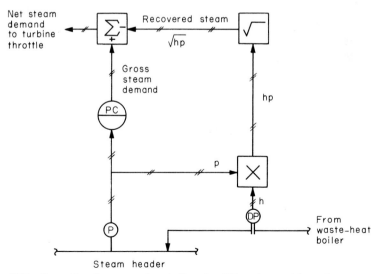

Fig. 3.18. Steam flow from waste-heat boilers should be subtracted from the output of the header-pressure controller to set the demand for steam supply to the header.

Controlling header pressures at their minimum acceptable values allows the recovery of more steam from waste-heat boilers. The rate of heat transfer through a boiler is proportional to the temperature difference between the process fluid and the boiling water. As boiler pressure is reduced, more heat can be transferred, which lowers the process temperature until a new equilibrium is established. Thus the process fluid leaves with a lower enthalpy, and more heat is recovered for use. Although the incremental heat recovery may be small when coupled with improved turbine power, reduced feedwater pumping power, etc., the savings due to minimizing header pressures and boiler pressure can be substantial.

2. Calculating Boiler Efficiency

In order to optimize a system of multiple boilers, it is necessary to determine the efficiency of each individual boiler as conditions change. Essentially all the heat losses in a boiler leave via the stack. They exist in two forms, however: unburned fuel, and sensible heat. Presumably, the combustion controls are adjusted to optimize the conversion of fuel to heat, such that the quantity of unburned fuel in the flue gas is negligible. In order to achieve this goal, however, excess air is required, whose percentage increases as load is reduced. As stated in Chapter 2, this adjustment is necessary because mixing of fuel and air is less complete when their velocities are low.

The increase in excess air usage at low loads causes an increase in flue gas losses, because that air not required for combustion is heated to the stack temperature and discarded. This impairs boiler efficiency at low loads.

To recover some of the heat in the flue gas, large boilers are fitted with economizers to heat feedwater, and air preheaters to warm the combustion air. These are convective devices because of the temperature ranges covered, and therefore follow the laws of heat transfer by convection. As the plant load increases, the rate of heat transfer from the flue gas to feedwater and incoming air must increase in direct proportion if heat loss is to remain a constant fraction of the fuel flow. However, the film coefficients on both sides of the heat-transfer surface increase only with the 0.8 power of velocity, such that they fail to keep pace with increasing flow rates. As a result, the temperature difference across the surfaces must increase somewhat with load, causing heat loss to thereby increase.

The heat lost to the stack is the sensible heat in the flue-gas mixture. Considering that the total mass of flue gas is the same as the mass of fuel and air entering the boiler,

$$Q_s = (F + A)C_p(T_s - T_a) + \Delta H_v F_w \qquad (3.16)$$

where F is the fuel flow (ash-free), A the air flow, C_p the average specific heat of the flue gas, and T_s and T_d the stack and ambient temperatures. The last term is the heat required to vaporize mass F_w of water formed in combustion. Airflow is set in ratio to fuel flow as

$$A = KF(1 + x) \qquad (3.17)$$

where K is the fuel–air mass ratio corresponding to zero excess air, and x is the fraction of excess air used in the combustion. The heat released by complete combustion of the fuel is

$$Q_c = FH_c \qquad (3.18)$$

where H_c is the ash-free higher heat of combustion of the fuel. Combining all of the above allows the formulation of the percent efficiency of a boiler:

$$
\begin{aligned}
E &= 100\left(1 - \frac{Q_s}{Q_c}\right) \\
&= 100\left\{1 - \frac{C_p[1 + K(1 + x)](T_s - T_a) - \Delta H_c}{H_c}\right\} \qquad (3.19)
\end{aligned}
$$

where ΔH_c, the difference between higher and lower heating values, takes into account the latent heat of water formed in combustion.

The efficiency of a given boiler is not constant, even at constant load and excess air. Soot accumulates and must be periodically removed. In addition, moisture entering with the fuel and air affects temperature distribution. Furthermore, the quantity of excess air being used at any time varies with the accuracy of the fuel and air flowmeters and heat of combustion of the fuel. Excess air can be controlled through analysis of the oxygen in the flue gas. Then x in (3.19) can be calculated from the mole-fraction oxygen in the flue gas, y:

$$x = K'y/(1 - y/0.21) \qquad (3.20)$$

where K' is 4.90 for coal, 5.09 for oil, and 5.26 for natural gas.

It should be noted that variations in the heat of combustion do not have a pronounced effect on efficiency. Actually, they are compensated by nearly equal changes in the air required. Therefore a small reduction in H_c will be almost completely offset by a comparable reduction in K. As a consequence, these two parameters may be considered constants for any given boiler with a single source of fuel.

In actual practice, there is relatively little difference in the constants for Eq. (3.19) whether firing a boiler with coal, oil, or gas, except for $\Delta H_c/H_c$. Reference (10) gives the theoretical air requirements for coal, oil, and natural gas as 7.85, 7.58, and 7.37 lb/10,000 Btu, respectively. Using these constants and calculating the specific heat of the flue gas based on its composition for each fuel reduces Eq. (3.19) to

$$E = 100\left[1 - 10^{-3}(0.22) + \frac{K''y}{1 - y/0.21}(T_s - T_a) - \frac{\Delta H_c}{H_c} \right] \quad (3.21)$$

where y is the mole fraction of oxygen in the flue gas, and K'' is a coefficient assigned to each fuel: 1.01 for coal, 1.03 for oil, and 1.07 for natural gas. The last term $\Delta H_c/H_c$ is about 0.02 for coal, 0.05 for oil, and 0.09 for gas.

This analysis has not considered the energy used by the boiler auxiliaries such as fans and pumps. They account for a small fraction of the energy input, and cannot vary appreciably with load. Boiler blowdown also represents a loss, but it is manageable and separately accountable. The principal loss is through the stack, and load balancing must be programmed to minimize its contribution.

3.   Multiple-Output Control Systems

When several boilers are sending steam to a common header, any one or combinations of more than one may be used to control header pressure. The very flexibility afforded by multiple boilers poses a decision-making problem. The operator and/or a computer must determine which boiler or boilers to use to control pressure. The next section describes a logical procedure for maximizing plant efficiency using a computer to make the decision. This section is devoted to developing a control system allowing the operator to do it.

Any control system manipulating several outputs to control a single variable should have certain features: It should

(a)   allow the operator to distribute the load,
(b)   prevent operator adjustments from upsetting control, and
(c)   control uniformly, regardless of the number of boilers in service, in automatic, or constrained.

The system shown in Fig. 3.19 accomplishes all these objectives.

The master pressure controller, with or without the assistance of feedforward, sets the needed total fuel flow. A total fuel flow controller manipulates all valves within the limits and distribution set by the operator. Should the operator change the distribution by adjusting a bias station ($\pm$), total fuel flow will change, causing its controller to readjust all individual valves to restore it to its previous level. Because flow measurements respond so quickly, the controller can normally correct for such an upset in less than 5 sec. Without this rapid correction, steam pressure would be upset, requiring minutes to recover.

It is also possible that the total-fuel controller may drive one of the valves to a limit; the controller will then simply drive the other valves until its set

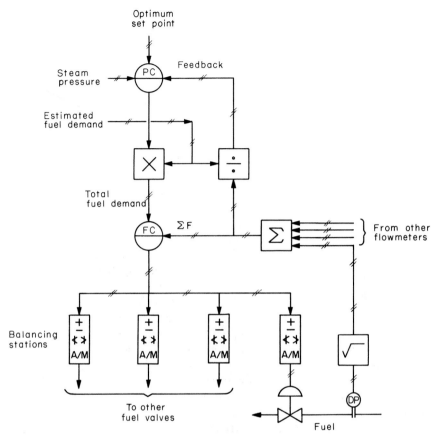

Fig. 3.19. In the multiple-output control system, a total-fuel controller manipulates all fuel valves, although the operator may distribute the load and set individual limits.

point is satisfied. When the number of boilers being driven by the total-fuel controller changes, its loop gain will change. Consequently, the total-fuel controller should be tuned with all boilers in automatic, so that loop gain can only decrease for all other situations and stability will be assured. Because the loop-gain variation is accepted by the total-fuel controller, the boiler-pressure loop has a constant gain. This is much preferable to a variable gain in the pressure loop because pressure is an order of magnitude more difficult to control than flow.

While Fig. 3.19 indicates bias (±) adjustments available to the operator, ratio adjustments may be preferable. Biasing allows the operator to elevate or lower the fuel flow to one boiler with respect to its fellows, with that difference being the same, irrespective of load. A *ratio* adjustments sets the percentage of total load accepted by a given boiler. Either station requires adjustable high and low limits, and automatic–manual transfer.

4. Maximizing Plant Efficiency

Given a number of boilers generating steam, the group efficiency can be increased if the most efficient boiler is allocated more of the load and the least efficient is allowed less. This strategy can be carried out on-line if a computer is available to calculate individual boiler efficiencies per Eq. (3.19). The computer then determines the most and least efficient boilers. On the next *increase* in load, only the most efficient boiler is manipulated to control total fuel flow; on the next *decrease*, only the least efficient boiler is manipulated.

Eventually the most efficient boiler will either enter a region of decreasing efficiency, where it will no longer be selected, or encounter a high limit. The computer must acknowledge the imposition of the limit and select the next most efficient boiler for future load increases. Similarly, the computer must pass over a boiler that has been placed in manual.

The least efficient boiler will only be increased in load when the others are either limited or in manual. However, it must accept all decreasing-load commands until it reaches a low limit. This may well force its efficiency lower. A boiler whose efficiency in the normal-load range is high may be less efficient than others at low load. This unit may have to be protected by a low limit or operator intervention, otherwise it might be shut down by the computer when it could contribute usefully, given a greater share of the load.

In the case where all boilers have identical characteristics, this system will typically drive some to their maximum efficiency, shut down others, and drive one to a load point either above or below maximum efficiency, depending on the load encountered. The selection is based on existing rather than projected efficiency, so that large load changes may move the plant away from the optimum, as the selected boiler is driven away from its peak. Nonetheless this action can be corrected by subsequent calculations and adjustments. It has the advantage of operating on actual measurements rather than on curves or models that may not be representative of true conditions.

Manipulation is achieved by using a separate total-fuel controller for each boiler as shown in Fig. 3.20. One of these controllers is in automatic at all times. Since each is manipulating its own valve and receiving a response only from its own flowmeter, its loop gain is unaffected by the condition of the rest of the plant. This system is therefore even more stable and responsive than that appearing in Fig. 3.19.

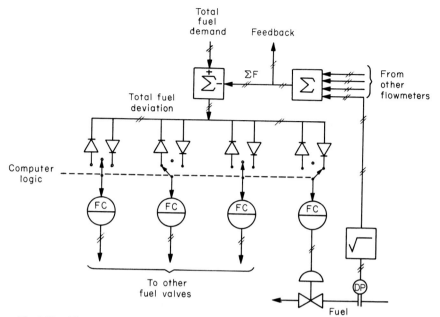

Fig. 3.20. The computer sets the switches to pass increasing demands (positive deviations) to the most efficient boiler and decreasing demands (negative deviations) to the least efficient boiler.

REFERENCES

1. Pollard, E. V., and K. A. Drewry, Estimating performance of automatic extraction turbines, General Electric Co., West Lynn, Massachusetts.
2. Nathan, P., and R. Wasserman, Energy control cuts electric bills, *Instrum. Technol.* (Dec. 1975).
3. Shinskey, F. G., Process control systems with variable structure, *Contr. Eng.* (Aug. 1974).
4. Shinskey, F. G., "Process Control Systems," McGraw-Hill, New York, 1967.
5. Yarway Steam Conditioning Valve, Yarway Corp., Blue Bell, Pennsylvania.
6. The Foxboro Company, Differential vapor pressure cell transmitter, Model 13VA, *Tech. Inform.* Sheet 37–91a (April 15, 1965).
7. Tomlinson, L. O., and R. W. Snyder, Optimization of STAG combined cycle plants, presented at the American Power Conference, Chicago, April 29–May 1, 1974.
8. Shinskey, F. G., Two-capacity variable period processes, *Instrum. Contr. Syst.* (Oct. 1972).
9. Shinskey, F. G., Adaptive control? Consider the alternatives, *Instrum. Contr. Syst.* (Aug. 1974).
10. Johnson, A. J., and G. H. Auth, "Fuels and Combustion Handbook," p. 355. McGraw-Hill, New York, 1951.

Chapter

4

Compressor Control Systems

Much of the energy used in petroleum refining and chemical processing powers compressors which feed reactors, provide refrigeration, or simply transport gaseous products. Because the energy used in compression is usually in the form of mechanical work, its conservation is of particular value. The only compressors not using mechanical energy are the ejectors which provide vacuum or compress low-pressure steam. But because they are compressors and use the work content of steam for their motive power, they are naturally included in this chapter.

The thermodynamic principles of compression are the same regardless of the type of compressor used—even steam jets must obey the rules. Therefore, the chapter begins with a development of the thermodynamics of compression before examining the properties of individual types of compressors. Then the demands of the process using the compressor are presented, which leads to the application of flow and pressure controls.

Special consideration is given to turbocompressors because of their extensive use in high-volume applications. Control systems must be specifically designed for each unit to minimize energy consumption while protecting the machine and its driver from abnormal operating conditions. These systems can become quite complex when multiple compressors or multiple streams around a single multistage compressor are encountered.

A. THE THERMODYNAMICS OF COMPRESSION

While isentropic compression was discussed in Chapter 1, the thermo-dynamic principles involved must be examined in more detail to estimate the effects of control action on compressor and process alike. Compression of ideal gases is considered first, to define parameters and evaluate efficiency factors. Then the relationships are applied to real gases and gaseous mixtures.

1. Adiabatic Compression of an Ideal Gas

If compression were carried out isothermally, the pressure–volume pro-duct of an ideal gas would be the same at both inlet and outlet conditions:

$$p_1 V_1 = p_2 V_2 = FRT/w \tag{4.1}$$

Here p is the absolute pressure, V the volumetric flow rate, F the mass flow rate, R the universal gas constant, T the absolute temperature, and w the molecular weight of the gas.

However, isothermal conditions cannot be maintained in most compres-sors due to the lack of sufficient heat-transfer surface; adiabatic conditions are more likely to be approached. Then the exponent of the pressure–volume relationship increases from unity (in the isothermal case) to the ratio of the specific heats of the gas, if the compression is carried out reversibly (isentro-pically):

$$p_2/p_1 = (V_1/V_2)^\gamma \tag{4.2}$$

where γ is the ratio of specific heat at constant pressure to that at constant volume, C_p/C_v. The absolute temperature of the discharged gas is elevated above that of the inlet in relation to the compression ratio

$$T_2/T_1 = (p_2/p_1)^\phi \tag{4.3}$$

where $\phi = 1 - 1/\gamma$.

In adiabatic compression, essentially all of the power introduced is converted into an increase in enthalpy of the fluid, regardless of the irrevers-ibility of the operation. For the flowing system, let W represent the power applied and F the mass flow of fluid. Then

$$-W = F \Delta H = F \int_{T_1}^{T_2} C \, dT \tag{4.4}$$

where C is the specific heat of the fluid and T its temperature. The applied power is also equatable as the product of mass flow and "adiabatic head" h_a, a parameter commonly used by compressor manufacturers:

$$-W = F h_a/\eta_a \tag{4.5}$$

Table 4.1

Molecular Weights and Specific-Heat Functions for Common Gases

Gas	Molecular weight, w	Specific-heat functions	
		$\gamma = C_p/C_v$	$\phi = 1 - 1/\gamma$
Air	29	1.40	0.286
Ammonia	17	1.31	0.237
n-Butane	58	1.09	0.083
Carbon dioxide	44	1.30	0.231
Carbon monoxide	28	1.40	0.286
Chlorine	71	1.36	0.265
Ethane	30	1.19	0.160
Ethylene	28	1.24	0.194
Hydrogen	2	1.41	0.291
Methane	16	1.31	0.237
Nitrogen	28	1.40	0.286
Oxygen	32	1.40	0.286
Propane	44	1.13	0.115
Propylene	42	1.15	0.130
Sulfur dioxide	64	1.24	0.194
Water	18	1.33	0.248

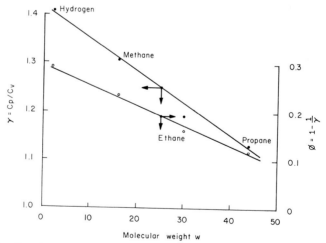

Fig. 4.1. Functions γ and ϕ vary nearly linearly with molecular weight for light hydrocarbons.

Here the adiabatic efficiency η_a indicates that not all of the power is converted into head. With flow expressed in pounds per minute and head in feet, power is presented in foot-pounds per minute.

Adiabatic head can then be calculated from temperature rise, using the conversion factor of 778.3 ft-lb/Btu:

$$h_a = 778.3\eta_a \int_{T_1}^{T_2} C\, dT \tag{4.6}$$

The enthalpy of an ideal gas is independent of pressure such that the specific heat used in (4.4) and (4.6) is that at constant pressure, C_p. Over the temperature rise usually encountered in compression, C_p may be considered constant, in which case (4.6) is reduced to

$$h_a = 778.3\eta_a C_p(T_2 - T_1) \tag{4.7}$$

Using the relationship

$$R = C_p - C_v = 1.987 \quad \text{Btu/lb-mol} \tag{4.8}$$

and substituting pressures for temperatures from (4.3) allows adiabatic head to be evaluated in terms of compression ratio p_2/p_1 at 100% efficiency. Then dividing by the molecular weight of the gas converts the pound-mole units to pounds, yielding adiabatic head in feet:

$$h_a = (1546 T_1/\phi w)[(p_2/p_1)^\phi - 1] \tag{4.9}$$

Table 4.1 lists the molecular weight and γ and ϕ functions for common gases. Those of the paraffin series and their mixtures follow a reasonably consistent relationship between these functions and molecular weight. Curves illustrating the relationship appear in Fig. 4.1.

2. Adiabatic and Polytropic Efficiencies

Adiabatic efficiency can be calculated by comparing the head produced to the corresponding temperature rise. Equations (4.7) and (4.9) are set equal, yielding

$$\eta_a = \frac{(p_2/p_1)^\phi - 1}{(T_2/T_1) - 1} \tag{4.10}$$

Equation (4.10) indicates that for any value of η_a less than 100%, Eqs. (4.2) and (4.3) no longer hold. In fact, they are valid only for isentropic compression, and must be corrected for real conditions of "polytropic" compression. The temperature–pressure relationship can be corrected simply by applying the polytropic efficiency factor η_p for the compressor:

$$T_2/T_1 = (p_2/p_1)^{\phi/\eta_p} \tag{4.11}$$

The advantage in using polytropic efficiency is that it is constant for a given compressor, regardless of the properties of the gas or the compression ratio. The Elliott Company (1), for example, lists the polytropic efficiency for its multistage compressors as being between 76 and 78%. Adiabatic efficiency can then be calculated from nominal polytropic efficiency by substituting (4.11) into (4.10):

$$\eta_a = \frac{(p_2/p_1)^\phi - 1}{(p_2/p_1)^{\phi/\eta_p} - 1} \tag{4.12}$$

Adiabatic efficiency will always be less than polytropic efficiency for any compressor. For a machine having η_p of 78%, compressing air ($\phi = 0.286$) adiabatically to a ratio of 5, η_a is calculated to be 72.7%. High compression ratios and high specific heat ratios lead to lower adiabatic efficiencies.

Polytropic head, occasionally used by compressor manufacturers, can be calculated from adiabatic head and the two efficiency factors:

$$h_p = h_a(\eta_p/\eta_a) \tag{4.13}$$

3. Real Gases

A pressure–enthalpy diagram can be used to compare inlet and outlet conditions for those fluids for which precise thermodynamic properties are available. First, the starting point is located on the diagram as in Fig. 4.2 from the intersection of inlet pressure and temperature coordinates.

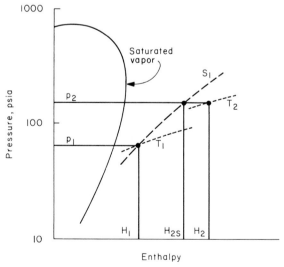

Fig. 4.2. The enthalpy increase following a line of constant entropy from p_1 to p_2 is divided by adiabatic efficiency to yield the actual enthalpy increase sustained by the compressed gas.

Then the contour of constant entropy S_1 is followed to the desired outlet pressure p_2. This locates the enthalpy H_{2S} the gas would be discharged at under isentropic conditions. Adiabatic head can then be calculated as

$$h_a = 778.3(H_{2S} - H_1) \qquad (4.14)$$

Given the polytropic efficiency of the compressor, the compression ratio, and the specific heat ratio for the gas, η_a can be calculated using Eq. (4.12).

The actual enthalpy H_2 at discharge is then calculated from the isentropic enthalpy increase divided by the adiabatic efficiency:

$$H_2 = H_1 + (H_{2S} - H_1)/\eta_a \qquad (4.15)$$

Locating H_2, p_2 on the diagram then defines the state of the discharged gas.

If a pressure–enthalpy diagram is unavailable, then the arithmetic average supercompressibility factor of inlet and outlet must be multiplied by Eq. (4.9) to determine the adiabatic head. Since the supercompressibility at outlet conditions may vary with outlet temperature, a second approximation may be necessary after determining outlet temperature from η_a by means of Eq. (4.7).

For mixtures of gases, ideal solutions can generally be assumed, where specific heats and their ratios are based on molar average values. Some compressor manufacturers such as Elliot (1) can generate thermodynamic properties of gaseous mixtures from computer models.

B.   COMPRESSOR CHARACTERISTICS

Adiabatic compression follows the preceding laws regardless of the nature of the compressor, although efficiencies may vary substantially from one type to another. However, each type of compressor has its own peculiar characteristics which determine, among other things, how it is to be controlled. Reciprocating compressors certainly have different control requirements than fans or steam-jet ejectors. In essence, the manipulable variables for each need to be enumerated, along with the constraints that must be obeyed to insure safe and efficient operation.

1.   Reciprocating Compressors

In a reciprocating pump or compressor, a certain volume (displacement) is swept from the suction to the discharge port with every rotation of the shaft. However, not all of the internal volume is displaced, i.e., a certain "clearance" remains undisplaced. Figure 4.3 shows one cylinder of a reciprocating compressor connected through valves to a suction line at p_1 and a discharge line at p_2. The piston is shown at bottom dead-center position, in

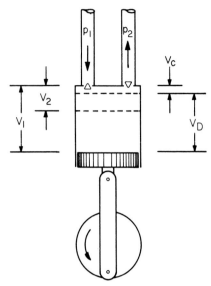

Fig. 4.3. The volume of gas contained at bottom dead-center is compressed to pressure p_2 before any discharge takes place; the piston then proceeds to top dead-center, leaving the clearance volume V_C.

which case the entire volume of the cylinder V_1 is filled with gas at suction conditions p_1, T_1.

As the piston is driven forward, a volume V_2 will be reached where the gas pressure equals the pressure p_2 in the discharge line. Only then can gas start to discharge. Using Eq. (4.2),

$$V_2 = V_1(p_1/p_2)^{1/\gamma} \qquad (4.16)$$

At top dead-center, volume V_D has been displaced, leaving clearance V_C remaining. The actual volume delivered during the stroke is $V_2 - V_C$. The "volumetric efficiency" of the compressor is the ratio of the delivered volume (at suction conditions) to the displacement:

$$\eta_V = [(V_2 - V_C)/V_D](p_2/p_1)^{1/\gamma} \qquad (4.17)$$

By substituting for V_2 from (4.16) and then $V_C + V_D$ for V_1, volumetric efficiency appears as a function of the clearance-to-displacement ratio and the compression ratio:

$$\eta_V = 1 - (V_C/V_D)[(p_2/p_1)^{1/\gamma} - 1] \qquad (4.18)$$

The ratio V_C/V_D can be expressed as the percent clearance. According to Evans (2), the actual volumetric efficiency will be 2–5% below theoretical for oil-lubricated compressors and 4–10% for nonlubricated designs, due to entrance losses and slippage.

The volumetric efficiency has no direct bearing on mechanical or work efficiency—it simply describes the capacity of the compressor that is actually in use compared to its nominal displacement. Many reciprocating compressors have "clearance pockets" that may be automatically opened to change their volumetric efficiency and therefore alter the rate of gas discharged. There is theoretically no loss in work efficiency in opening these pockets, although entrance and exit losses through valves will, in practice, have an effect.

Volumetric flow at suction conditions is the product of displacement, volumetric efficiency, and rotational speed. Then a plot of volumetric efficiency versus compression ratio is equivalent to a plot of volumetric flow versus compression ratio for a given speed. Figure 4.4 is a solution of Eq. (4.18) for air, allowing for 5% slippage, illustrating the effects of compression ratio and clearance on delivered flow.

In addition to clearance pockets, suction valves may be automatically "unloaded", i.e., held open through the entire cycle. This effectively removes a cylinder from service without stopping rotation of the shaft. Some losses are sustained in that flow passes through the suction valve in both directions. But where control cycles are short, the energy and wear associated with

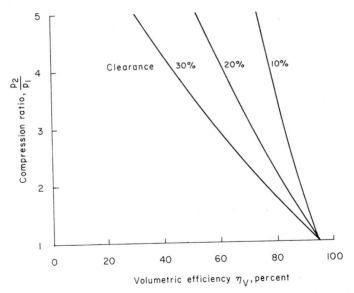

Fig. 4.4. The effect of clearance pockets on volumetric efficiency of a reciprocating compressor increases with compression ratio.

starting and stopping the compressor are avoided. Unloaders are used in single-cylinder compressors, providing on–off operation between two pressure limits. When multiple cylinders are working in parallel, unloaders alter the capacity of the unit stepwise, much like clearance pockets but without the additional influence of compression ratio.

Both unloaders and clearance pockets give stepwise control over capacity. If constant pressure or flow is necessary, other variables must be manipulated. Speed manipulation is most desirable but is limited to those units driven by steam or combustion engines or turbines. Variable-speed electric drives are generally too costly for this type of service.

Throttling valves on the suction or discharge of positive displacement compressors should be avoided—damage can result from a closed valve. Instead, a portion of the discharged flow is usually returned to the suction to control suction pressure, discharge pressure, or flow. The recycle stream should be taken downstream of an aftercooler, however; if the heat of compression is not removed before recycling, the temperature of the recycled gas will continue to rise and could cause damage. A linear valve is recommended.

2.   Rotary Compressors

Rotary compressors are positive-displacement devices without valves, where the gas is moved from inlet to outlet by sliding vanes, gears, or meshing lobes. Another design uses a liquid seal to minimize wear when handling dirty gases. Rotary compressors resemble reciprocating units in requiring either speed manipulation or a recycle valve for pressure or flow control— they cannot be dead-headed.

Rotary compressors are used primarily for pressures near or below atmospheric—vacuum service is quite common. Monroe (3) compares lobe and liquid-ring compressors to blowers and steam-jet ejectors for vacuum service. His curves of adiabatic efficiency versus suction pressure follow Eq. (4.12) with constant polytropic efficiency up to a point. For example, the liquid-ring compressor has a polytropic efficiency in the range of 35–39% up to 200 mm Hg, where it begins to fall; its adiabatic efficiency peaks at 28% about 250 mm. The lobe compressor has a polytropic efficiency of 55% up to 60 mm Hg; its adiabatic efficiency reaches 42% at 120 mm, falling somewhat at higher pressures.

3.   Turbocompressors

Turbocompressors are dynamic devices, creating a head by imparting momentum from an impeller to the gas. Their two principal styles are centrifugal and axial. The centrifugal compressor operates much like a centrifugal pump, the fluid being given a radial thrust to the wall of the casing where it is withdrawn. An axial compressor drives the gas parallel to the shaft through a series of turbine blades as in a jet engine. The two compressor styles have substantially different flow characteristics as shown in Figs. 4.5 and 4.6.

The centrifugal compressor is best applied where a constant head is required, whereas an axial unit is more suitable where a constant flow is needed under conditions of variable pressure. The flow range of the centrifugal compressor exceeds that of the axial, but both are limited by a "surge region," an area of unstable operation entered as flow is reduced.

In the surge region, the head-flow characteristic of a turbocompressor actually reverses slope, developing a negative-resistance characteristic. As flow is reduced, discharge pressure falls, causing flow and pressure to be further reduced. When discharge pressure falls below that in the line, a momentary flow reversal occurs, and line pressure starts to fall. This creates a demand for more flow, causing flow to reverse again. These pulsations will continue until control action is applied to force the compressor out of the surge region, or until damage develops.

Adequate flow must be provided to keep a compressor out of surge when the process demand is low. This flow may be provided by recirculating

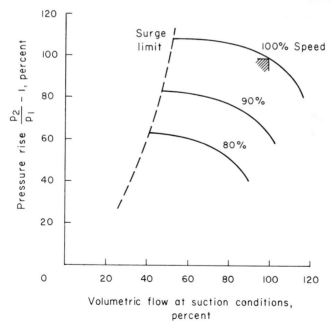

Fig. 4.5. Head-flow characteristics for a centrifugal compressor.

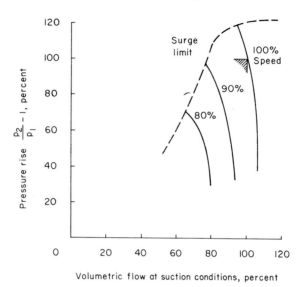

Fig. 4.6.   Head-flow characteristics for an axial compressor.

cooled discharge gas back to the suction through a bypass control valve. For an air compressor having atmospheric suction, the additional flow may be provided by simply venting enough compressed air directly to the atmosphere. However, bypassing and venting waste energy, so every effort should be taken to minimize these practices.

A variable-speed driver extends the operating range of a compressor considerably, and should be used when wide ranges in operating conditions are anticipated. Generally speaking, flow varies linearly with speed for a fixed head. However, in the turbulent regime, head varies with flow squared, hence with speed squared. This relationship limits the speed range of a compressor to about its upper 20 or 30%.

If the adiabatic efficiency of a compressor were constant, power would vary simply with the product of flow and head. In light of the flow-speed and head-speed relationships already described, power would then vary with speed cubed. However, adiabatic efficiency decreases somewhat at higher compression ratios per Eq. (4.12), altering the cubic relationship somewhat. The actual power used as a function of process load depends on the nature of the process and its control requirements, e.g., constant pressure, constant flow, or some combination of the two.

If speed is not variable, then throttling of the suction is normally used for control, preferably with inlet guide vanes. The vanes not only reduce suction pressure, and therefore flow and/or discharge pressure, but they also apply prerotation to the gas. A simple suction valve does not provide this, and consequently is not as efficient a throttling means. Characteristic curves for a constant-speed centrifugal compressor with adjustable guide vanes are given in Fig. 4.7.

Evans (2, p. 59) describes a centrifugal compressor with guide-vane throttling at 70% flow as using about 72% power, whereas suction-valve throttling requires about 75% power. By contrast, speed manipulation requires only about 68% power for the same load. Comparing Figs. 4.5 and 4.7 indicates that guide-vane manipulation seems to extend the flow range beyond what is available with speed variation. However, actual operating ranges are sensitive to process parameters as well. Generally speaking, variable-speed is preferred to save energy at reduced loads, with guide-vane throttling being a second choice.

Axial compressors are available with adjustable stator vanes between stages of rotor blades (4). They effectively double the turndown achievable with speed control alone. The surge limit moves toward lower flow and head as stator vanes are closed, much in the same way as with guide vanes for a centrifugal compressor.

Throttling the suction of a compressor reduces flow by reducing suction pressure. Note that the abscissa for compressor characteristic curves is given in units of volumetric flow at inlet conditions. Closing the suction valve will not change inlet volumetric flow, but will reduce the mass flow being compressed.

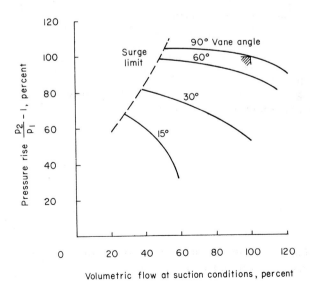

Fig. 4.7. Manipulation of inlet guide vanes of a centrifugal compressor allows turndown approaching speed manipulation.

Consider the compressor of Fig. 4.7 operating at the surge limit with a vane position of $30°$: the pressure rise would be 82.6% and the volumetric flow 37.7%. The compression ratio would then be

$$p_2/p_1 = 1 + 0.826p_r$$

where p_r would be the rated pressure rise for the machine. If suction valve throttling were used instead, the compressor characteristic would be represented by the single curve at $90°$ vane angle. Minimum flow for this curve occurs at 104% rated pressure rise, such that

$$p_2/p_s = 1 + 1.04p_r$$

where p_s is the compressor suction pressure downstream of the throttle valve. Suction pressure corresponding to the desired pressure rise of 82.6% can be found by equating the two above expressions:

$$p_s/p_1 = (1 + 0.826p_r)/(1 + 1.04p_r)$$

Then the minimum flow for suction throttling referred to nominal inlet pressure p_1 is

$$V_{m1} = 50(p_s/p_1) = 50(1 + 0.826p_r)/(1 + 1.04p_r)$$

For the compressor of Fig. 4.7, the rated suction and discharge pressures are 12.1 and 18.6 psia, respectively, giving

$$p_r = 18.6/12.1 - 1 = 0.537$$

Then

$$V_{m1} = 50\frac{1 + 0.826(0.537)}{1 + 1.04(0.537)} = 46.3\%$$

This example demonstrates that suction-valve throttling does not afford the extended range the guide-vane provides.

Throttling of the *discharge* valve allows no turndown at all from the maximum capacity curve. Therefore, the minimum flow attainable for the example cited would be 50%, compared to 46.3% for suction throttling and 37.7% for guide-vane throttling. Furthermore, the power required for discharge throttling is greater than that required for the other means. As a result, discharge throttling is rarely used to control compressors.

4. Ejectors

Ejectors are compressors which utilize a high-pressure fluid as the motive force. For example, high-pressure steam is commonly used to pump air from a vacuum chamber to atmospheric pressure. The compression ratio for a single stage is limited to about $10:1$ for reasonable efficiency. Ejectors may be arranged in series, with condensers between if the fluid is condensible. If the motive fluid is steam, the size and steam requirements of downstream stages are reduced by adding interstage condensers.

Steam-jet ejectors are very popular for providing evacuation of process vessels, down to as low an absolute pressure as 10 μ. They are attractive both because of a low installed cost and little required maintenance. However, their efficiency is particularly low. Monroe (3) reports thermal efficiencies as low as 1–3%, based on the heat content of the motive steam above that of boiler feedwater. This measure of efficiency would seem justifiable if the heat content of the exhaust steam were not recoverable. However, there are other meaningful ways to report efficiency.

The adiabatic efficiency of a compressor was defined by Eq. (4.5). Characteristic curves for steam jets are available in Refs. (2, p. 87) and (5), giving the "entrainment ratio" of steam-jet ejectors in terms of the mass of air or gas compressed per unit mass of steam or motive gas, as a function of suction, motive, and discharge pressures. The adiabatic efficiency for each case can be calculated from the adiabatic head of the gas compressed and the available work in the motive fluid. Assuming that the available work is entirely lost because the discharged mixture must be vented, or condensed and vented, adiabatic efficiency calculated in this way is comparable to that estimated for positive displacement pumps earlier.

Evans (2, p. 87) gives the steam–air mass ratio as a function of absolute suction pressure for one to six stream jets arranged in series with intercondensers. The curves generally overlap, forming an envelope whose edge is roughly approximated by a straight line on log–log coordinates. This line, representing the maximum efficiency for the optimum combinations, is represented by the equation

$$\log p_1 = 2.88 - 2.21 \log(F_S/F) \tag{4.19}$$

where p_1 is expressed in Torr, and F_S/F is the steam–air mass ratio; 2.88 is $\log 760$ Torr; steam is supplied at 100 psig. Table 4.2 lists the adiabatic

Table 4.2

Maximum Efficiencies of Multi-Stage Ejectors Using 100 psig Steam

Suction pressure p_1 (Torr)	Steam–air ratio F_S/F	Adiabatic efficiency η_a (%)	Polytropic efficiency η_p (%)
250	1.65	7.8	18.1
100	2.50	10.8	39.1
10	7.10	11.8	40.2
1	20.1	9.7	46.4
0.1	57	7.2	50.0

efficiencies calculated using Eqs. (4.5) and (4.19), with 363 Btu/lb taken as
the available work of saturated steam at 100 psig.

A single configuration loses adiabatic efficiency rapidly as the limits of its
range are approached. The efficiency of a single-stage steam-jet ejector is
given in Table 4.3 from points taken from one of the curves in Ref. (2, p. 87).

Table 4.3

Efficiencies of a Single-Stage Ejector Using 100 psig Steam

Suction pressure p_1 (Torr)	Steam–air ratio F_S/F	Adiabatic efficiency η_a (%)	Polytropic efficiency η_p (%)
500	0.65	6.7	11.3
250	1.65	7.8	18.1
100	6.0	4.5	19.9
50	60	0.67	17.7

Polytropic efficiencies for the multistage and single-stage ejectors were
calculated from η_a by solving Eq. (4.12) for η_p:

$$\eta_p = \frac{\phi \log(p_2/p_1)}{\log[1 + ((p_2/p_1)^\phi - 1)/\eta_a]} \qquad (4.20)$$

On the basis of the polytropic efficiencies shown in Tables 4.2 and 4.3, multistage ejectors can be as efficient as vacuum pumps at low absolute pressures,
but the single-stage ejector is less than half as efficient.

Increasing motive steam pressure improves the adiabatic efficiency of an
ejector. When the steam pressure for the combination giving 100 Torr suction
pressure in Table 4.2 is increased to 150 psig, η_a increases to 11.5%, while
falling to 8.5% when steam pressure is reduced to 70 psig.

Steam jets are often used to compress boiloff from evaporators enough to
allow further extraction of heat. In this service, adiabatic head and efficiency
have little significance—heat transfer is the objective. Reference (6) describes
a steam-jet compressor supplied with 474 lb/h of 150 psig saturated steam,
compressing 526 lb/h of water vapor removed from an evaporator at 7.7 psia,
to produce 1000 lb/h at 0 psig and 242°F for heat to the evaporator. No
energy is lost in the mixing process, yet available work is lost. Dividing the
available work of the discharge by the sum of that in the motive and suction
streams yields a work efficiency of 84.5%. Adiabatic efficiency under the
same conditions is estimated to be only 7.6%.

The efficiency parameter that is of most significance to the evaporation
process is the improvement in steam economy brought about by the com-

pressor. As a result of the difference in latent heat between condensing steam and evaporating water, along with heat losses, 1000 lb of steam removes only 850 lb of vapor. With the jet compressor, 474 lb of steam at 150 psig removes the same amount of vapor, increasing the vapor–steam ratio of the unit from 0.85 to 1.79. In essence, double-effect evaporation is obtained with a single vapor body. Increasing the steam pressure to 250 psig increases the vapor–steam ratio to 1.88, although the work efficiency is lower at 81.9%.

Ejectors are designed to be operated at a single steam flow, resulting from a stipulated pressure applied to a fixed nozzle. Control over pressure in the vessel being exhausted is effected by either throttling the suction line, bleeding air into the suction line, or manipulating the heat-transfer rate of an upstream condenser where condensible vapors are present. The problem of finding the optimum combination of condensing rate is discussed in relation to steam-jet refrigeration and evaporators in subsequent chapters.

Neither suction throttling nor air bleed affect steam flow—consequently the jet consumes the maximum flow of steam regardless of plant load. However, the velocity of the mixture flowing through the ejector varies with suction pressure. In certain regions of flow and pressure, a given ejector may exhibit instability (7). Figure 4.8 describes a characteristic curve that has been observed in some ejectors.

As long as the suction pressure of the ejector in Fig. 4.8 is below 10 Torr or above 30 Torr, pressure responds monotonically to the suction valve admitting air. But should the pressure rise above 10 Torr, the pumping rate falls until nearly 100 Torr is reached, causing the pressure to rise spontaneously to that point. A similar behavior is observed when attempting to reduce pressure below 30 Torr. An attempt to control in this region will

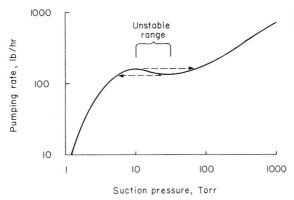

Fig. 4.8. This steam-jet ejector has a range of negative resistance, in which pressure cannot be stably controlled.

result in constant-amplitude oscillations, called "limit cycling." The only solution is to reconfigure the jets to provide stable characteristics throughout the operating range.

Reference (2, p. 87) gives a steam-usage multiplier, provided to correct the ejector performance curves for steam motive pressures other than 100 psig. Table 4.4 summarizes the effect of variable steam pressure on usage and adiabatic efficiency.

Table 4.4

The Effect of Steam Pressure on Usage, Capacity, and Efficiency at 250 Torr

Steam pressure (psig)	Steam–air ratio	Steam flow multiplier	Airflow multiplier	Adiabatic efficiency (%)
70	2.19	0.74	0.56	6.2
100	1.65	1.00	1.00	7.8
150	1.45	1.43	1.63	8.4

Because the nozzle operates at sonic velocity (the absolute upstream pressure being more than double the absolute downstream pressure), steam flow is directly proportional to absolute upstream pressure. The efficiency also varies with motive pressure, with the resultant airflow being a combination of the two effects. The only way steam flow to a single ejector can be reduced is to reduce the motive pressure. Alternately, multiple ejectors can be arranged in parallel, with some being valved off under conditions of reduced load.

Reference (8) describes a jet compressor used to raise the pressure of a stripped gas to that of the fuel-gas header, with the motive force being a high-pressure purge stream being let down to the same header. Pressure controllers on both gas streams throttle valves at their respective inlets to the compressor.

C. PRESSURE AND FLOW CONTROLS

Compressors may be used in a variety of services. They are used to supply feedstocks to reactors and to exhaust gaseous products from reactors. They also transport gases, both at the source and as boosters along the pipeline. Another common service is as heat pumps for evaporators, distillation columns, and refrigeration units. Each of these processes has its own particular properties and needs. Some require flow control, others pressure control, and the relationship between flow and pressure tends to vary from one application to another. This section considers process characteristics and demands, and how they interact with the characteristics of the compressor.

1. Line and Load Characteristics

The simplest case to consider is that of an air compressor drawing atmospheric suction and discharging into a process having both a static head and a dynamic head caused by resistance to flow. Figure 4.9 describes such an arrangement. The static head could be developed by a head of liquid or by a pressure-control loop as shown—the effect is the same. In some cases the static head is constant or atmospheric—in others it can vary, based on process requirements.

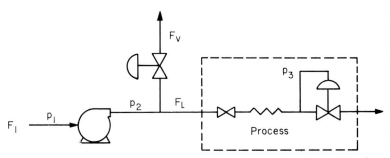

Fig. 4.9. The air compressor delivers into static pressure p_3 plus a dynamic head $p_2 - p_3$ developed by resistance to flow through the process.

Turbulent flow is encountered almost universally, such that the dynamic head varies with mass flow F_L as

$$F_L = k_L \sqrt{p_2(p_2 - p_3)w/T_2} \tag{4.21}$$

where k_L is a constant representing resistance to flow, pressures are in absolute units, w is the molecular weight of the gas, and T_2 is its absolute temperature at p_2. Mass flow F_L is proportional to volumetric flow at suction conditions:

$$V_{L1} = F_L R T_1/wp_1 \tag{4.22}$$

Combining the two expressions allows volumetric suction flow representing the process load to be plotted against compression ratio p_2/p_1 on the same graph as the compressor characteristics:

$$V_{L1} = k_L(RT_1/p_1)\sqrt{p_2(p_2 - p_3)/wT_2} \tag{4.23}$$

Resistance factor k_L must be expected to vary, particularly where there are many users of the compressed gas. Some typical load lines developed using (4.23) are plotted on compressor coordinates in Fig. 4.10.

For a constant load, the steady-state relationship between pressure and flow resulting from manipulation of compressor speed or vane position is determined by the slope of the load line. Conversely, when the load changes at constant speed, pressure and flow will vary along a contour of constant speed. This relationship bears directly on the stability of surge-control systems, which use the vent valve of Fig. 4.9 (or an equivalent recirculation valve) to prevent surge by changing the load.

Refrigeration compressors are used to pump heat from a cold process to an atmospheric reservoir or to a hot stream. When the heat load is zero, a temperature difference still normally exists between the hot and cold sides of the system, equivalent to a static head. The compressor suction pressure will tend to approach the vapor pressure of the refrigerant at the controlled temperature of the process, while its discharge pressure will approach the vapor pressure at the temperature of the stream being heated. As the heat load increases, both the cold-side and hot-side exchangers will develop proportional temperature gradients, causing the compressor suction pressure to fall and its discharge pressure to rise. Consequently, heat pumps also encounter a process load line. For a given unit, the static head will vary with the temperatures of source and sink; the slope of the load line may also be altered by temperature controllers which throttle, flood, or bypass the heat exchangers.

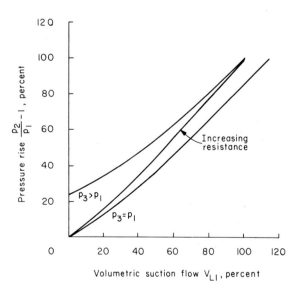

Fig. 4.10. The process resistance and static head may both be subject to change.

2. Balancing Supply and Demand

There are essentially two operating modes for compressors—constant flow and constant pressure. The constant-flow mode is typified by an axial compressor supplying air to a blast furnace. The mass flow of air to the furnace sets its production rate and therefore must be controlled. However, the resistance the furnace presents to airflow tends to change with time, developing fluctuations in head at a given flow. The superior flow-regulating characteristic of an axial compressor resists changes in head, yet speed and/or stator-vane manipulation are still required, particularly where production rate is variable.

A compressor sending air or gas to a number of parallel users is likely to be a centrifugal type. It is important to hold a constant pressure on the discharge header so that each individual user may be insured an adequate supply. The static head is likely to remain constant, but the slope of the load line will vary with demand. Compressor speed, vane, or suction-valve position must be manipulated to hold a constant discharge pressure.

Most of these systems provide a constant discharge pressure regardless of whether it is actually needed. A user reduces its flow simply by throttling from the constant supply pressure. If the number of users is low, the compressor discharge pressure can be minimized by a valve-position controller acting on the most-open valve, as shown in Fig. 4.11. Each valve is under control individually by its user; the valve-position controller provides that the header pressure is only high enough to satisfy the most demanding user.

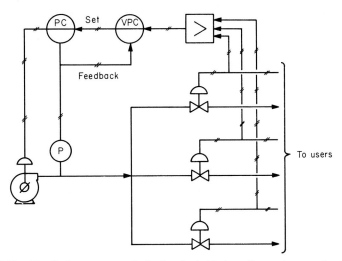

Fig. 4.11. The discharge pressure is slowly adjusted to keep the most-open valve in a nearly wide-open position.

The valve-position controller must be slower acting than all the user controllers—otherwise it will upset users by changing pressure too rapidly. If the users are under flow control, the flowmeters may require compensation for variable header pressure.

If there are many users, or if they are remote, the optimum discharge pressure may be correlated with flow. If the system in Fig. 4.11 were to supply but one user, the action of the VPC would convert the load from one of variable resistance to constant resistance. With a constant supply pressure, control over flow to the user can only be achieved by varying the resistance of the control valve. But by returning the control valve to the same position following changes in flow, the VPC would regulate the load resistance. With a constant resistance, pressure follows flow along a single load line.

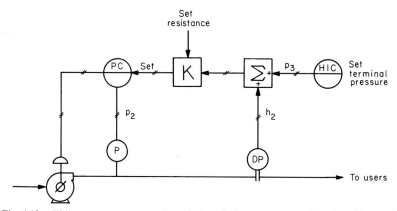

Fig. 4.12. This system saves energy by reducing discharge pressure with reduced flow, while maintaining a constant terminal pressure.

Then a constant-resistance control system can be built as in Fig. 4.12, wherein the pressure set point is driven proportional to measured flow. The required relationship between flow and pressure appears as in Eq. (4.21). However, if an orifice meter is to be used, its differential pressure must be related to flow:

$$F_L = k_m \sqrt{h_2 p_2 w / T_2} \qquad (4.24)$$

where k_m is the meter factor and h_2 is the differential pressure developed across it. When (4.24) is set equal to (4.21), flowmeter differential can be related to pipeline or process differential pressure:

$$p_2 - p_3 = h_2 (k_m / k_L)^2 \qquad (4.25)$$

In actual practice, this system has been used at water pumping stations, to maintain a desired static head p_3 at the user location, by adjusting discharge pressure p_2 for line losses at varying loads. It is equally applicable to compressors serving multiple users at a distant location.

3. Protecting Process and Driver

Satisfying the demands for flow and pressure is but one facet of controlling a compressor. During normal operation, nothing further is required. But abnormal conditions are to be expected, and process, compressor, and driver must be protected against damage should they develop. Compressor protection is a complex issue and is treated in detail in the next section. At this point, consideration is given to simpler systems for protecting the process from overpressure and the driver from overload.

Only one of the three interdependent variables—flow, suction pressure, and discharge pressure—may be regulated by manipulating the compressor. Should another require control, another manipulated variable must be found, such as another stream entering or leaving the system. However, in most cases, control over other variables need only be exercised in the limit, e.g., to avoid an overpressure.

Consider a centrifugal compressor whose discharge pressure is ordinarily controlled by throttling the suction valve. At low loads, the valve might be throttled to the point where suction pressure falls below atmospheric, thereby drawing gland-sealing oil into the compressor. If this is a constraint, then a suction-pressure controller should override the discharge-pressure controller whenever suction pressure falls to that point. When an override takes place, discharge-pressure control is lost, and pressure will rise above its set point—but this is preferable to drawing oil into the compressor.

At the other extreme, it may be possible to open the suction valve wide enough to overload the driving motor. Should an overload occur, the circuit breaker would then trip, and the compressor would be lost altogether, requiring an operator to bring it back on line. To avoid this possibility, a motor-load controller can override the suction valve at a point just below the setting of the circuit breakers. Again, pressure deviation will develop, but a deviation is preferable to losing the service of the compressor altogether.

The two override controllers appear in conjunction with the discharge pressure controller in Fig. 4.13. The high selector prevents the suction valve from closing entirely, while the low selector stops it from opening to the point of overload. Feedback of the valve position to all three controllers will insure that an override will be effected with little, if any, overshoot.

Overload is frequently encountered when the compressor is started. However, startup conditions vary significantly between single- and multiple-compressor installations, and are complicated by antisurge controls. For these reasons, a discussion of startup considerations is included under the heading of multiple-compressor installations.

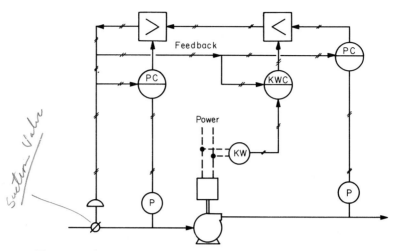

Fig. 4.13. The kilowatt controller can close the valve to prevent motor overload, and the suction-pressure controller can open the valve to avoid loss of seal oil.

D. SURGE PROTECTION

The most demanding aspect of controlling compressors is surge protection. The problem lies in being unable to determine with absolute certainty the degree of approach to surge. Once a compressor begins to surge, it will continue until corrective action is applied, so automatic protection is mandatory. A small centrifugal may surge many times without damage, but a 100,000-hp axial could require reblading after a single incident.

When a compressor begins to surge, the suction flow falls to zero within a few milliseconds, reverses momentarily, and begins to recover in less than a half second. If the situation is not corrected, the cycle repeats immediately, resulting in a series of thunderclaps less than a second apart. The sudden fall in suction flow can be detected and used to open a recirculation valve, but not before at least one surge cycle is sustained. To prevent surge from developing at all requires a control system which skirts the unstable area altogether.

1. Outlining the Surge Region

The compressor characteristic curves represent a roadmap, with the area to be avoided clearly delineated. Using a set of process measurements, the actual operating condition at any point in time should be determinable, along with the degree of approach to surge. But two difficulties appear:

(1) The true characteristics of the compressor may deviate from its reported characteristics due to wear, variable gas composition, etc.

(2) The related variables of adiabatic or polytropic head and volumetric suction flow are not directly measurable.

Strictly speaking, curves relating pressure rise against suction flow are valid only for a single set of suction conditions, i.e., pressure, temperature, and composition. Particularly when composition is variable, compressor manufacturers will use adiabatic or polytropic head in place of pressure rise. This, then, describes the properties of the machine apart from its service. Presuming these properties are constant, it remains to the control engineer to outline the surge region in terms of measurements which he has available.

White (9) observes that the surge line appears as a parabola when adiabatic head is plotted against volumetric suction flow:

$$h_a = kV_1^2 \qquad (4.26)$$

There are exceptions to this rule, particularly with regard to axial compressors, where the surge line often turns parallel to the baseline at high flow rates (see Fig. 4.6).

However, volumetric flow cannot easily be measured directly, but is usually detected as orifice differential h_1:

$$V_1 = k_m \sqrt{h_1 T_1 / p_1 w} \qquad (4.27)$$

where k_m is a meter factor. When Eq. (4.27) is substituted into (4.26) and set equal to (4.9), a relationship between orifice differential and suction and discharge pressures forms:

$$kk_m^2 \frac{h_1 T_1}{p_1 w} = \frac{1545 T_1}{\phi w}\left[\left(\frac{p_2}{p_1}\right)^\phi - 1\right]$$

Observe that temperature and molecular weight terms cancel, leaving h_1 in terms of p_2, p_1, and ϕ:

$$h_1 = k_s (p_1/\phi)[(p_2/p_1)^\phi - 1] \qquad (4.28)$$

The supercompressibility factor for real gases has been omitted from both Eqs. (4.9) and (4.26), but it would cancel in the same way as T_1 and w.

The relationship between h_1 and $p_2 - p_1$ would be linear if ϕ were unity, but unfortunately it is far from unity. Consequently, there is a substantial departure from linearity for all but the lowest compression ratios. To illustrate, Fig. 4.14 plots the bracketed term of (4.28) against percent pressure rise for three ranges of compression ratios for air. Departure from linearity increases from about 9% with a maximum compression of 3 to about 25% when the maximum compression ratio is 50.

For compression ratios below 2, the relationship is nearly linear, and its slope can be estimated by differentiating the function:

$$d(p_2/p_1)^\phi / d(p_2/p_1) = \phi(p_2/p_1)^{\phi-1} \qquad (4.29)$$

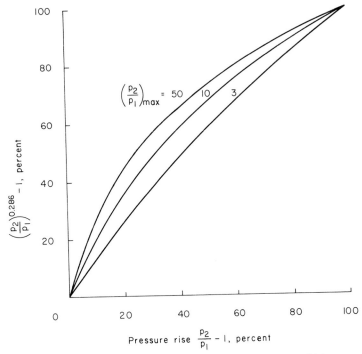

Fig. 4.14.   The function of pressure rise shows increasing nonlinearity at higher compression ratios.

Evaluated at a compression ratio of unity, Eq. (4.29) becomes ϕ, in which case

$$(p_2/p_1)^\phi - 1 \simeq \phi[(p_2/p_1) - 1] \tag{4.30}$$

When (4.30) is substituted into (4.28), the approximate relationship becomes simply

$$h_1 \simeq k_s(p_2 - p_1) \tag{4.31}$$

While accurate only for very low compression ratios, (4.31) has the advantage of being independent of composition and temperature.

2.   Antisurge Control Systems

The most common antisurge control systems are based on the approximation of Eq. (4.31). Orifice differential h_1 is controlled in ratio k_c to measured compressor differential Δp, with k_c adjusted to be a safe margin above the calculated surge ratio k_s. The system is shown in Fig. 4.15. Opening the

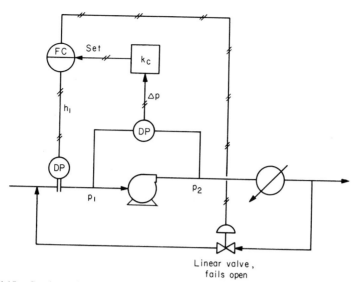

Fig. 4.15.   Suction orifice differential is held in ratio to compressor differential by manipulating the recirculation valve; a pressure or flow controller manipulating speed or a suction valve has been omitted for simplicity.

recirculation valve (or vent valve as in Fig. 4.9) increases h_1 and reduces Δp at the same time, thereby forcing the compressor away from surge.

In actual practice, the recirculation or vent valve changes the load on the compressor. When it is opened, pressure and flow follow a line of constant speed, which is generally perpendicular to the surge line, and therefore the compressor is moved directly away from the surge region. To protect the compressor from surge in event of a control failure, this valve should open on loss of signal, power, or air supply.

The valve must also be capable of opening quickly, 1- or 2-sec stroking time being common. In addition, it should have a linear rather than equal-percentage characteristic. In most cases, Δp does not change greatly, so an equal-percentage characteristic is not needed. In fact, it is undesirable, because an equal-percentage valve passes much less flow than a comparably sized linear valve at all positions except wide open.

Recirculation or venting must be recognized as an abnormal condition in that it wastes energy. Consequently, the antisurge valve is nearly always closed, and the compressor is nearly always operated away from the surge region. In this condition, suction flow will be well above that required for surge control. The controller, in attempting to reduce the flow, drives its output up to the full limit of its supply voltage or pressure. Prolonged operation in this state results in windup of the integral mode, which retards

control action when it is next needed. Should the load suddenly decrease, a
conventional controller cannot open the antisurge valve until its deviation
crosses zero, in which case the compressor may temporarily go into surge.

Integral windup can be inhibited by a circuit which attempts to hold the
controller output at the valve limit by manipulating internal feedback (10).
The controller output then lingers at the closed valve position rather than
driving to full supply voltage or pressure. And as soon as its deviation from
set point starts to move toward zero, the valve starts to open. Consequently
the valve opens before zero deviation is reached, preventing surge from
developing even temporarily. Should the load not change sufficiently to
force the deviation to zero, the valve will slowly return to a closed position.
In effect, the proportional action of the controller is always available for
surge protection, while integral action is enabled only when the valve is not
closed.

3. Improving System Accuracy

The linear approximation to the surge curve as represented by Eq. (4.31)
is limited in accuracy in several respects. First, the actual relationship between
head and flow may not be parabolic as was assumed. This is particularly
true for axial compressors and for those equipped with movable vanes.
Reference (4) describes a surge envelope of the type shown in Fig. 4.16 when
plotted on coordinates of Δp and h_1. In addition to the linear approximation
already developed, this compressor requires a limit on Δp; however, both
the slope of the control line and its limit must change with vane position.

A control system which provides the necessary functions is shown in Fig.
4.17. The system solves the equation

$$\Delta p = (k_0 + k_h\theta)h_1 < (p_0 - k_p\theta) \tag{4.32}$$

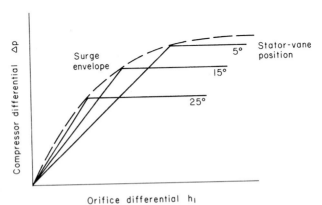

Fig. 4.16. Each stator-vane position requires a different pair of control lines.

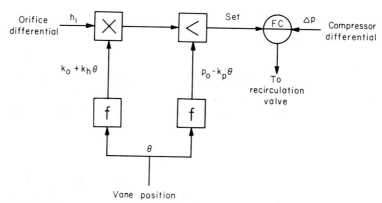

Fig. 4.17. This system provides a two-section control line which moves as a function of vane position.

where k_0 is the slope of the control line at zero vane position (θ) and p_0 is the limit of $p_2 - p_1$ under those same conditions; k_h and k_p represent the variations in slope and limit as a function of θ.

The linear model also fails to match the actual surge curve even at moderate compression ratios, as Fig. 4.14 demonstrates. Consequently, points taken from compressor characteristic curves plotted on coordinates of Δp and h_1 should not be expected to fall in a straight line, even if the relationship between head and flow is exactly parabolic. When Eq. (4.28) is solved for conditions of constant suction pressure with the results plotted as Δp versus h_1, a single curve appears. However, each suction pressure gives a new curve, so that p_1 must be included in the controls if the model is to be exact.

Another approach is to model the appropriate curve of Fig. 4.14 with a straight line of the form

$$(1/\phi)[(p_2/p_1)^\phi - 1] \simeq a(p_2/p_1) + b \tag{4.33}$$

Coefficients a and b would be adjusted for the best "safe" fit, i.e., so that the line would lie tangent to, but above, the curve, and balanced across the expected operating range. When (4.33) is substituted into (4.28), we have

$$h_1 = k_s(ap_2 + bp_1) \tag{4.34}$$

Again, coefficients a and b would be selected by the designer, but k_s would be a ratio setting adjustable by the operator to correct any observed mismatch between model and compressor.

Variations in composition that change ϕ will have some effect on the approximation of (4.33). At moderate compression ratios, however, the effect is not severe. For example, at a compression ratio of 5, a change in heat-capacity ratio from 1.40 to 1.30 (reducing ϕ from 0.286 to 0.231) changes the functional group represented by (4.33) from 2.04 to 1.95—only a 5% reduction. The magnitude of the error increases to about 7% at a compression ratio of 10.

By contrast, the simpler system based on the linear relationship between h_1 and Δp is sensitive to variations in molecular weight and temperature as well as ϕ and p_1. White (9) describes the need to compensate for temperature and composition variations where they become significant, when h_1 and Δp are used.

4. Using Available Measurements

All of the antisurge control systems just described require a flowmeter installed in the suction of the compressor. But in many situations, piping constraints rule against this location. Frequently, a discharge flow measurement will be used as a substitute, but it must be corrected to suction conditions. Because the same mass flows through both meters, their differential pressures would be related as

$$h_1 = h_2(p_2 T_1/p_1 T_2) \tag{4.35}$$

It is actually not necessary to correct for the temperature ratio, because it is related to the compression ratio per Eq. (4.11). Then (4.35) reduces to

$$h_1 = h_2(p_2/p_1)^{(1 - \phi/\eta_\mathrm{p})} \tag{4.36}$$

Table 4.5 summarizes this relationship for air compressed with a polytropic efficiency of 78%.

Table 4.5

The Discharge Flowmeter Correction Factor for Air at a Polytropic Efficiency of 78%

Compression ratio p_2/p_1	Correction factor $(p_2/p_1)^{(1 - \phi/\eta_\mathrm{p})}$	Compression ratio p_2/p_1	Correction factor $(p_2/p_1)^{(1 - \phi/\eta_\mathrm{p})}$
1.2	1.12	10	4.30
1.5	1.29	20	6.67
2	1.55	50	11.91
5	2.77		

As the table indicates, the correction factor is always much lower than the compression ratio, mitigating to some extent the need for compensation. A modified square-root extractor could be used to raise the compression ratio approximately to the power needed for compensation.

Occasionally, neither suction nor discharge flowmeter are available. Then a speed measurement may be used to develop a surge line as a function of

p_2 and p_1 or Δp. When a constant-speed motor drives the compressor, its power consumption or vane position can be used to relate to Δp at the surge limit. Evans (2, p. 56) shows contours of constant horsepower intersecting the surge limit on a plot of pressure rise versus suction flow for both constant-speed and variable-speed compressors.

5. Interaction with Pressure and Flow Control

During normal operation, pressure is controlled by manipulating speed, vane, or suction-valve position. When the surge line is approached, however, the surge controller is activated, opening the vent or recirculating valve. Motion of this valve acts fundamentally as a load change on the compressor, upsetting the controlled pressure. However, the head-flow curves for a centrifugal compressor tend to be flat as the surge line is approached. Then changes in load caused by the antisurge controller will move the compressor away from the surge line without greatly affecting pressure.

Conversely, with a constant output from the antisurge controller, changes in compressor speed, vane, or suction-valve position tend to move along a line of constant load. Figure 4.18 describes the paths followed for a variable-speed air compressor with atmospheric suction and resistive loading (i.e., $p_3 = p_1$). The antisurge controller opens the vent valve, increasing flow and dropping pressure slightly. The pressure controller would react with a slight increase in speed, which also moves the compressor away from the surge

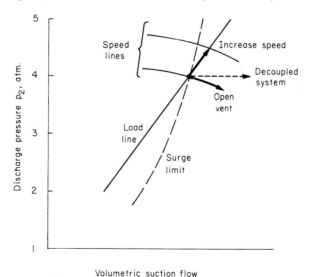

Fig. 4.18. Opening the vent valve decreases pressure along a speed contour, while increasing speed increases pressure along a load line.

line. These two loops tend to operate stably with little interaction, but only in that given configuration.

If the pressure controller were connected to the vent valve and the antisurge controller to the governor, stability would be sacrificed. In this configuration, opening the vent valve will lower pressure but only *temporarily*. The antisurge controller would then attempt to return the compressor to the surge control line by increasing speed, which would *raise* pressure more than the vent valve lowered it. Consequently the pressure response to a step in vent valve position would be a small drop followed by a large rise. This characteristic is known as inverse response and is singularly responsible for many failures in multivariable control system (11). Furthermore, when the vent valve is closed, as in normal operation, pressure control must be transferred to speed manipulation, unnecessarily complicating the control system.

The same argument can be made for controlling flow into the process load because, at constant load, flow is proportional to discharge pressure.

A greater degree of interaction between pressure and antisurge control might be expected with higher static loads (reducing the slope of the load line) or steeper speed curves. Then the vent valve may have more effect on pressure, and speed may influence the surge calculation to a greater degree. Decoupling can be provided to minimize this natural interaction as shown in Fig. 4.19. As the antisurge controller opens the vent valve, compressor speed is increased proportionately. If the proper weighting is placed on the two vectors in Fig. 4.18, their resultant would be horizontal, i.e., pressure would remain constant. The negative sign on the vent valve input to the summer is required because the valve opens on a decreasing signal.

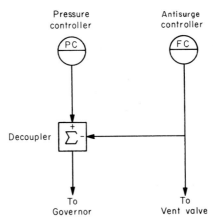

Fig. 4.19. Opening of the vent valve is automatically converted into a proportional increase in speed to avoid upsetting pressure.

E. MULTIPLE-COMPRESSOR INSTALLATIONS

Compressors are not often used singly. Most installations require two or three in parallel to extend the operating range, to allow for maintenance, startup, and abnormal conditions. Some compressors are connected in series with intercooling to achieve high compression ratios efficiently. They may be mounted on a single shaft, or be driven independently, as determined by the needs of the process using the gas.

1. Series Arrangements

Compressors connected in series on the same shaft are controlled essentially in the same manner as a single multistage machine. Multistage compressors are characterized by a composite surge curve, although series compressors on the same shaft may not be, particularly if some stages may be bypassed, or if flows may be extracted or injected between compressors. In most cases, a single antisurge control system is sufficient, using a flowmeter on the suction of the first stage.

If other stages carry different flows from the first due to extraction or injection, additional calculations and measurements are required. Last-stage suction flow may be inferred from discharge by Eq. (4.36). Interior suction flow rates must be determined by mass balances of streams entering and leaving. The resulting mass flow rates representing interstage suction flows must be converted into volumetric flow or equivalent orifice differentials for use in surge control.

Series compressors driven by different shafts require completely separate antisurge systems. However, it is possible to use a single recirculation valve that bypasses both compressors. In fact, if individual valves are provided, opening of one valve for surge protection will almost certainly cause the other to open. Two antisurge controllers can operate a single valve through a low selector—whichever sends a decreasing signal to open the valve will be the one taking over control. Both controllers require the antiwindup feature described earlier.

2. Parallel Arrangements

Parallel compressors require some special considerations. Each must have its own antisurge control system, because there is no guarantee that the load will be shared equally between them. Their characteristics may not necessarily be identical due to age and wear, and in many cases parallel compressors will be different models. Then a change in load may affect one more than another, such that surge protection must be available to all. Load balancing is discussed in more detail in the next section.

While it is possible to protect all compressors in a parallel installation with a single recirculation or vent valve, it is not recommended. Individual compressors are isolated from one another with check valves as shown in Fig. 4.20. This prevents backflow into a shutdown compressor. It is then not possible to start up a compressor connected in parallel with others already in operation unless it has its own recirculation or vent valve. The operating compressors have already established a head, which the idle compressor must overcome before it can deliver any flow. If started into an established differential, it will immediately surge.

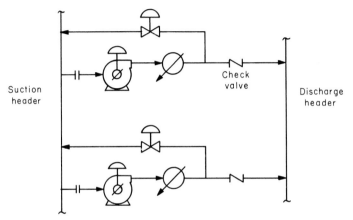

Fig. 4.20.   Parallel compressors are isolated from each other with check valves.

If each compressor is provided with its own dedicated antisurge system, it cannot surge upon startup. As speed increases, its head will rise until it overcomes the established differential and begins delivering its share of flow. The antisurge controller will then begin to close the recirculation valve.

At the opposite extreme, a compressor motor may become overloaded if started with the recirculation valve fully open. In this case, it may be necessary to position the recirculation valve manually to a point where both overload and surge are avoided. Then once operating speed is reached, the antisurge controller can be transferred to automatic, where it will begin closing the valve and enabling the compressor to take up the load.

When one of a parallel group of compressors enters the surge-control mode, its recirculation valve will begin to open. Because *any* recirculation valve in fact bypasses *all* operating compressors, the recirculated flow is distributed among all. Then much more flow is recirculated than needs to be, to keep the troubled compressor out of surge. This is grossly inefficient, and is a signal that the operating load is too low to justify the number of compressors on line. At this point, one of the compressors should be automatically shut down. It is not necessary to shut down the one operating under antisurge control—the release of any compressor will increase the load to those remaining, taking them away from the surge line. To eliminate the effects of transient disturbances, shutdown should only be initiated after a reasonable time delay.

Automatic startup and shutdown of parallel compressors can be accomplished by the system shown in Fig. 4.21. Should any compressor reach its maximum allowable guide-vane position or motor load, an alarm unit starts another compressor. The appropriate compressor is selected through a rotating sequencer—a system of logic which chooses the compressor which has been idle longest for startup, and the one which has been operating longest for shutdown. Any compressor may be withdrawn from the sequence manually.

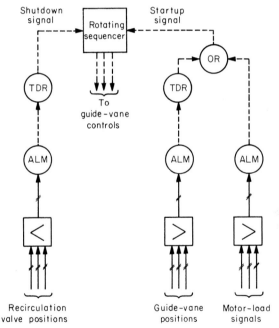

Fig. 4.21.   A compressor is automatically shut down when a recirculation valve opens, and one is started when any guide-vane or motor load reaches its maximum allowable value.

3.   Balancing for Maximum Efficiency

Slight variations in speed or characteristic curves can easily cause a maldistribution in load when parallel compressor operate into the same head. Suction flowmeters can be used to provide a balance. Since the compressors may not all be the same size or have identical characteristics, the total load must be distributed on a adjustable percentage basis. The total flow, as measured independently or as summed from individual flowmeters as in Fig. 4.22, sets all ratio flow controllers (FFC). Each seeks to maintain its own flow in the set proportion to the total by slowly biasing its vane position (or speed) as required. The operator must judge how to set the ratios to maximize the efficiency of the entire unit. Alternately, a computer could optimize the load distribution, as done for boilers in Fig. 3.20, after calculating individual efficiencies on the basis of delivered flow divided by

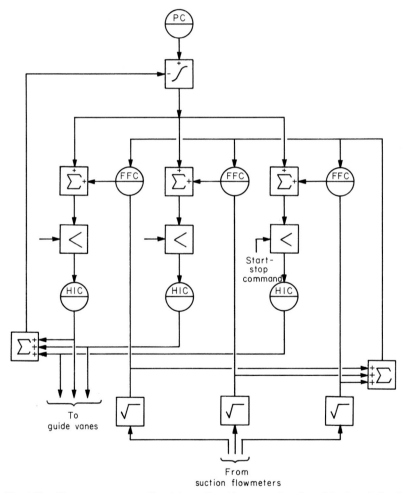

Fig. 4.22. The pressure controller drives all guide vanes directly, while flow balancing controllers bias individual vanes slowly.

motive power. Since all compressors are operating into the same head, pressures may be left out of the calculation.

It is imperative that the pressure controller act directly on the vanes (or governor) without being cascaded through a flow controller. The response of the flow loop is close enough to that of the pressure loop to interfere with its performance. In fact, systems designed for pressure-flow cascade control of compressors have had to be changed in the field to direct pressure control, to achieve the desired degree of stability and response.

In any system where a single controller must manipulate several outputs in parallel, however, a feedback loop must be added to protect the controlled variable against upsets and gain changes when units are brought on and off line. This feature was also provided for boilers and generators in Chapter 3. In Fig. 4.22, the output of the pressure controller is sent to summing units where flow balancing is applied, then through low selectors for start and stop commands, and finally through auto–manual stations (HIC) before

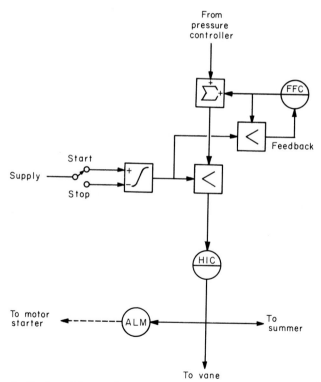

Fig. 4.23. Application of the supply signal to the positive input of the integrator will ramp open the vane to the point where pressure control takes over; switching the signal to the negative input ramps the vane closed, at which point the motor stopped.

connection to the guide vanes. The actual vane-position signals are summed and fed back to a high-speed integrator at the output of the pressure controller. This integrator assures that the sum of the vane positions always corresponds to the output of the pressure controller, regardless of how many compressors are on line, or how their loading is balanced. Because there are only instruments in this feedback loop, it can be made to respond in a fraction of a second by appropriately adjusting the integrator, and will not therefore degrade the response of the pressure-control loop.

The start–stop circuit for each compressor is detailed in Fig. 4.23. The guide vane is ramped open or closed by applying a supply voltage (or pressure) to the integrator. In the "start" position of the switch, the integrator will ramp the vane open until the pressure controller assumes manipulation; the integrator output continues to rise to maximum so that it will not interfere with control. In the "stop" position, the integrator drives downward from its maximum output, requiring a little time to override the control signal. Eventually it will close the vane, and the motor may be stopped.

The act of opening or closing a guide vane is fed back to the integrator in the pressure control circuit, automatically adjusting the other vanes for the disturbance. Notice also that the ratio flow controller is protected against windup by taking its feedback signal through a selector whose other input is the ramp.

REFERENCES

1. Elliott Division, Carrier Corp., Elliott Multistage Compressors, Bulletin P-25, Jeannette, Pennsylvania, 1973.
2. Evans, F. L., Jr., "Equipment Design Handbook for Refineries and Chemical Plants," Vol. 1, p. 76. Gulf Publishing Co., Houston, 1971.
3. Monroe, E. S., Energy conservation and vacuum pumps, *Chem. Eng. Progr,* (Oct. 1975).
4. Fallin, H. K., and J. J. Belas, Controls for an axial turboblower, *Instrum. Technol.* (May 1968).
5. Perry, R. H., and C. H. Chilton, "Chemical Engineers Handbook," 5th ed., p. 6–31. McGraw-Hill, New York, 1973.
6. Economical Evaporator Designs by Unitech, *Bulletin* UT-110, Ecodyne Unitech Division, Union, New Jersey, 1976.
7. Shinskey, F. G., Controlling unstable processes, Part I: The steam jet, *Instrum. Contr. Syst.* (Dec. 1974).
8. Shah, B. M., Saving energy with jet compressors, *Chem. Eng.* (July 7, 1975).
9. White, M. H., Surge control for centrifugal compressors, *Chem. Eng.* (Dec. 25, 1972).
10. Shinskey, F. G., Effective control for automatic startup and plant protection, *Can. Contr. Instrum.* (April 1973).
11. Shinskey, F. G., Interaction between control loops, Part II: Negative coupling, *Instrum. Contr. Syst.* (June 1976).

Chapter

5

Refrigeration

Refrigeration is used extensively not only in industrial processing but increasingly to air-condition buildings. Most refrigeration systems are driven by mechanical compressors, using electricity or high-pressure steam or gas turbines for power. Hence, they are work-intensive and tend to be quite costly to operate. They therefore present considerable opportunity for energy conservation.

Additionally, all refrigeration systems ultimately pump heat into the environment—either into the atmosphere or cooling water. As a consequence, their performance is quite sensitive to atmospheric conditions. In most cases, control systems are applied to isolate refrigeration systems from atmospheric disturbances, which then prevents them from taking advantage of favorable conditions. The loss in performance thus sustained is doubly damaging when the principal component of heat load is leakage to the atmosphere, since it is affected by the same conditions.

This chapter will describe common refrigeration systems and the factors affecting their performance, culminating in controls to maximize that performance under whatever conditions prevail.

A. MECHANICAL REFRIGERATION

1. Single-Stage Units

A simple mechanical refrigeration system is described in Fig. 5.1. The compressor draws evaporating refrigerant from the chiller where heat is removed from the process stream. Compressed to an elevated pressure and temperature, the refrigerant is condensed against atmospheric cooling. The condensed liquid then is drawn through an expansion valve where it is partially flashed to vapor at the lower pressure of the chiller. The remaining liquid boils at that pressure to complete the cycle.

The controls shown in Fig. 5.1 are not coordinated with each other, and are responsible for considerable inefficiency in the heat-pump cycle, as conditions change. An investigation into the factors determining the performance of the unit will reveal the problem areas and point the way to a more efficient control system.

The amount of work W required to remove an amount of heat Q_c from a body at a temperature T_c into a heat sink at temperature T_s in a Carnot cycle is

$$-W/Q_c = (T_s - T_c)/T_c \qquad (5.1)$$

where all temperatures are given in absolute units. Recognize, however, that the Carnot cycle is based on isentropic compression and expansion, along with reversible isothermal heat transfer. In practice, these conditions cannot be approached, such that the actual work required will always be substan-

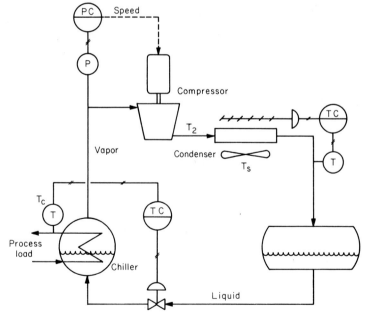

Fig. 5.1.   These uncoordinated control loops prevent the refrigeration unit from operating at maximum efficiency.

tially higher than that given in Eq. (5.1). Nonetheless, the simplicity of the equation is useful to demonstrate the profound effect of source–sink temperature difference on the performance of a heat pump. Control systems should be designed to minimize that temperature difference insofar as possible, consistent with ambient conditions and heat load.

A simple refrigeration cycle is shown on a pressure–enthalpy diagram in Fig. 5.2. Refrigerant evaporating at T_1 and p_1 is compressed to T_2 and p_2 along a curve deviating somewhat from an isentropic contour. The superheated vapor is then cooled and condensed at pressure p_2. Throttling through the expansion valve delivers to the evaporator a mixture having the enthalpy of saturated liquid at p_2.

The heat absorbed per unit refrigerant in the evaporator is $H_1 - H_0$. The work consumed in compression is $H_2 - H_1$. The work required to transfer a unit of heat is then

$$-W/Q_c = (H_2 - H_1)/(H_1 - H_0) \qquad (5.2)$$

If the refrigerant were an ideal gas, Eq. (5.2) could be solved using the familiar parameters of specific heats and latent heats, without referring to a diagram. Compression work could be found from heat capacity C_p,

$$-W = C_p(T_2 - T_1) \qquad (5.3)$$

and discharge temperature could be found from the compression ratio,

$$T_2/T_1 = p_2^{\phi/\eta_p}/p_1 \qquad (5.4)$$

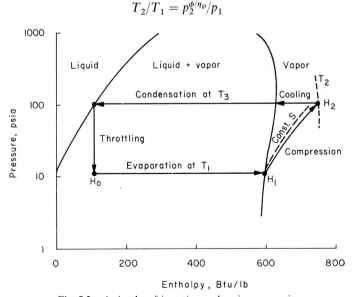

Enthalpy, Btu/lb

Fig. 5.2. A simple refrigeration cycle using ammonia.

where $\phi = 1 - C_v/C_p$ and η_p is the polytropic efficiency of the compressor. However, refrigerants tend to deviate from ideality, so caution should be exercised in using this approach.

Once a particular cycle has been described and evaluated it is possible to equate the true work–heat ratio to the Carnot expression by applying an appropriate efficiency factor:

$$- W/Q_c = (T_3 - T_1)/\eta_c T_1 \tag{5.5}$$

Having evaluated the cycle efficiency η_c for a nominal set of conditions, Eq. (5.5) may then be used to investigate the influence of temperatures on the work–heat ratio. Recognize however, that the condensing and evaporating temperatures T_3 and T_1 depart from sink and source temperatures T_s and T_c by the temperature differences across the two heat-transfer surfaces. At zero load, $T_1 = T_c$ and $T_3 = T_s$; as the heat load increases, the temperature differences will increase:

$$T_1 = T_c - Q_c/U_c A_c \tag{5.6}$$

$$T_3 = T_s + (Q_c - W)/U_s A_s \tag{5.7}$$

where U and A are the effective overall heat-transfer coefficient and area for the heat exchanger designated by the subscripts.

The dominant factor in the heat–work relationship is the temperature difference $T_3 - T_1$. To conserve energy, this temperature difference should be minimized insofar as possible. Where it is partially contingent on heat transfer, that heat transfer should be facilitated by maximizing $U_c A_c$ and $U_s A_s$—in general, the controls applied to refrigeration units, typified by Fig. 5.1, do not. By attempting to control condenser temperature (or pressure) at a constant value, the controller inhibits heat transfer, raising $T_3 - T_s$, and therefore T_3, above its minimum value. Similarly, T_c is controlled by the expansion valve, which holds the refrigerant level in the chiller below maximum, thereby reducing the effective area of heat transfer. As a result, $T_c - T_1$ is higher than necessary, and therefore T_1 is lower than it would be without controls. In fact T_1 is held arbitrarily at the boiling point of the refrigerant at the controlled pressure in the chiller. This is neither necessary nor desirable.

The three control loops are completely uncoordinated. Each is designed to function assuming constant values for the other two controlled variables. Minimizing $T_3 - T_1$ requires that heat transfer through both heat exchangers be unrestricted. When the heat load Q_c and ambient temperature T_s change, then suction and discharge pressures and temperatures should also change—this is the only way to minimize the work required for compression.

Figure 5.3 shows the refrigeration unit with controls rearranged to reduce compression work. Control over condenser temperature is eliminated to take

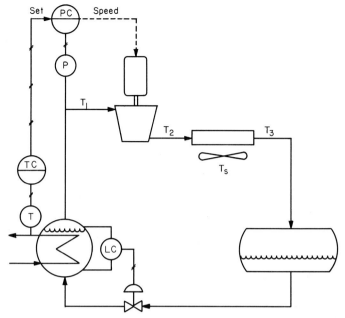

Fig. 5.3. Rearrangement of the controls to use all available heat-transfer surface maximizes cycle efficiency.

full advantage of ambient variations. In addition, level in the chiller is controlled at a point where all tubes are immersed, and heat transfer is therefore not restricted by dry surface. This allows a higher pressure in the chiller for a given controlled temperature and heat load. The pressure controller will then reduce speed, and with it, suction flow and horsepower.

Lacking control over condenser temperature will cause substantial variations in it as ambient changes develop. The cycle will become more efficient at night, during rainy weather, and in the winter. At some point a constraint may arise (such as the surge limit of the compressor) beyond which no further savings in power are realizable. Also worth considering are savings in fan horsepower possible during periods of low ambient temperatures. A fan driven by a two-speed motor can be reduced in speed under these conditions; or in a multiple-fan installation, some fans may be turned off. These operations are much simpler and less costly than the incorporation of louvers as in Fig. 5.1 or variable-pitch blades. While these adjustments are necessary if a uniform temperature or pressure is required, they are superfluous for the optimization problem just described.

Ideally, expansion of refrigerant liquid to the pressure of the chiller would be conducted through an engine rather than a valve. Figure 1.10 shows that it results in less vaporization as well as recovering some of the work of compression. Unfortunately, an engine operating on saturated liquid faces severe mechanical stresses. The liquid is easily propelled in slugs by the rapidly expanding vapor, causing extensive erosion.

Instead, where pressure differences are large, the refrigerant is expanded in two or three stages through valves. At each stage, vapor is removed and returned to an appropriate point between compressor stages. This helps reduce superheat within the compressor, and reduces suction flow, both of which improve cycle efficiency.

2.   Serving Multiple Users

Multistage compressors are often used to supply refrigeration to users at several temperatures by means of "side-loading." It is based on the same multistage-flash principle described above. Liquid refrigerant is available at each stage of flashing, evaporating at the corresponding stage of compression, as shown in Fig. 5.4.

The control system is arranged to provide only enough compressor power to meet the needs of the most-open user valve. Like other valve-position controllers, this is set for slow manipulation of compressor speed to avoid upsetting those users not selected for control. There is, however, one distinc-

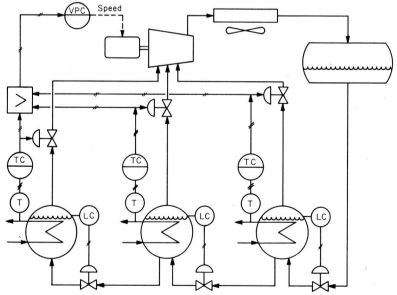

Fig. 5.4.   The multistage compressor provides refrigeration at several temperatures with a single fluid.

tion between this flow pattern, which is serial, and parallel patterns. If one of the user valves is consistently much more open than the others, it may indicate an evaporating pressure that is too high for the heat load. Reconnection of that user to a lower-pressure stage could provide a better balance among all users. Ideally, all control valves should approach the full-open position, where work loss due to throttling is minimized.

Chillers connected in parallel, using the same liquid and evaporating into the same vapor header, can be controlled in a similar fashion. However there is another arrangement that eliminates all vapor valves and is therefore worth considering—it is shown in Fig. 5.5. Rather than throttling the vapor, liquid is throttled to reduce the tube surface exposed to boiling, thereby regulating heat transfer. The chiller requiring the most heat transfer will have the highest level of refrigerant. Then compressor speed can be used to control the highest refrigerant level to achieve the same effect as the vapor valve-position controller used earlier.

The major drawback of the variable-level chiller is its poor response to control action. In order to change heat-transfer rate, level must change first; and the rate of level change is limited by the capacity of the chiller and the rate of boiloff. Consequently, response of temperature to refrigerant valve position is slow. But there is an additional complication: boiling refrigerant tends to foam, wetting more tube surface than is indicated by a liquid weight

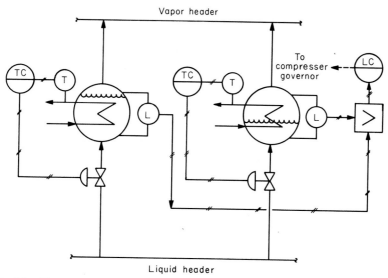

Fig. 5.5. The compressor must be controlled to satisfy the most demanding user, which is the chiller with the highest level of refrigerant.

measurement. In fact, all the tube surface may be wet when the weight measurement indicates the chiller to be only half full. The degree of foaming is proportional to the heat load, which provides a certain degree of self-regulation in response to load changes. However, response to control action attempting to correct a deviation in temperature may not be linear or reproducible. Then it is difficult to find controller settings which are both responsive and stable for all operating conditions. In an effort to provide stability, responsiveness often must be sacrificed.

3. Precise Temperature Control

Occasionally, extreme precision is required over process temperature, to provide a maximum amount of chilling without freezing, as an example. This requires maximum responsiveness, so that liquid level must be held constant and evaporating pressure manipulated for control. The rate of heat transfer will be directly proportional to the temperature difference between the process fluid and the evaporating refrigerant. Therefore, for a given controlled temperature, the refrigerant temperature, as determined by its pressure, must change with heat load in a manner indicated in Fig. 5.6.

As the heat-transfer rate approaches zero, there will be no gradient across the tube surface, and the refrigerant temperature will be at the control point. To increase the heat-transfer rate, refrigerant temperature must fall from that point in a reasonably linear fashion. Over the ranges of operating conditions encountered in a system such as this, the vapor pressure of a refrigerant is

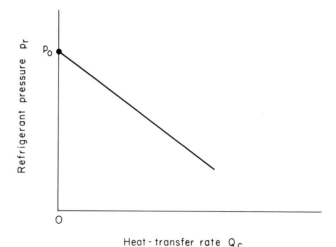

Fig. 5.6.   Refrigerant pressure must decrease as heat load increases to maintain a constant process temperature.

usually linear with temperature and is therefore generally linear with heat-transfer rate. Where this assumption is not justified, an estimate of refrigerant pressure at two or three conditions other than zero heat load can be used to provide a more representative curve.

Figure 5.7 shows how a feedforward control system would be implemented, based on this relationship. The heat load is calculated from process flow F and the desired reduction in temperature:

$$Q_c = FC(T_i - T_o^*) \tag{5.8}$$

where T_o^* is the desired outlet temperature of the process fluid, T_i is its measured inlet temperature, and C is its heat capacity. (Note: the superscript asterisk is used to distinguish set points from measured variables.) Feedback trim is customarily applied in the location where the controller set point appears in the feedforward equation (1). Then the output m of the feedback controller appears in place of T_o^* in Fig. 5.7. Having estimated the heat-transfer rate, it is then necessary to program the refrigerant temperature T_r proportional to it:

$$T_o^* - T_r^* = KQ_c \tag{5.9}$$

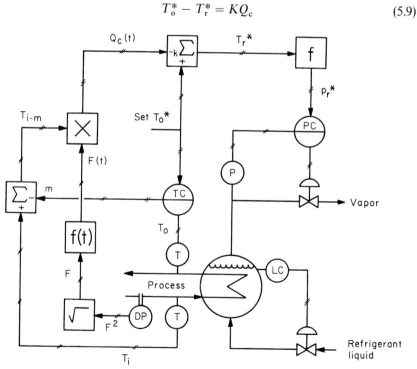

Fig. 5.7. This feedforward system sets refrigerant pressure proportional to calculated heat load.

where K is an adjustable constant representing heat-transfer area and coefficient. Finally the desired refrigerant pressure p_r^* is extracted as a function of its temperature:

$$p_r^* = f(T_r^*) = f[T_o^* - KF(t)C(T_i - m)] \qquad (5.10)$$

Dynamic compensation in the form of a first-order lead-lag is usually necessary to match the response of process temperature to refrigerant pressure, to that resulting from changes in flow F. Its presence is indicated in Eq. (5.10) by the functional (t). See Ref. (1, pp. 211–219) for details on its contribution and adjustment.

An important feature of the system is its ability to return refrigerant pressure consistently to the vapor pressure corresponding to the control point, upon loss of heat load. This protects against freezing on loss of load, regardless of whatever action may have been taken by the feedback controller in the course of operation. As zero load is approached, the feedback controller loses its influence over the process.

Coefficient K should be adjusted to match the slope of the line in Fig. 5.6 to the true process characteristic. If inlet temperature T_i is not significantly variable, it may be omitted from the calculation, and the output of the feedback controller may be connected directly into the multiplier. The equation thus developed is

$$p_r^* = f[T_o^* - mF(t)] \qquad (5.11)$$

System complexity is thereby reduced, and the feedback controller takes over the task of slope adjustment.

A valve-position control system similar to that shown in Fig. 5.4 can be used to conserve compressor power.

4. Compressor Load Lines

Load lines for processes using compressed gases were presented in Chapter 4 to locate operating points and speculate on the effects of control action. Similar lines can be drawn for refrigeration systems by converting heat flow into volumetric suction flow, and temperature difference into head.

If suction pressure is constant, mass flow and volumetric flow are equivalent. Also if the pressure in the chiller is constant, the latent heat of evaporation will be constant as well. Then volumetric suction flow is proportional to heat flow, making the two scales equivalent. However, an example follows, where the actual variation in suction pressure is taken into account in establishing the load line.

Adiabatic head was found in Chapter 4 to be related to temperature rise through the adiabatic efficiency factor:

$$h_a = 778.3\eta_a C_p(T_2 - T_1) \qquad (5.12)$$

However adiabatic efficiency η_a is not constant, varying with compression ratio. Fortunately, there is a direct substitution of polytropic head and efficiency for adiabatic head and efficiency:

$$h_p = 778.3\eta_p C_p(T_2 - T_1) \tag{5.13}$$

It will be recalled that η_p is a property of the compressor alone, being independent of compression ratio and gas composition. For a refrigerant of constant heat capacity, then, polytropic head is linear with temperature rise.

A few other variables need to be evaluated before any conclusions can be drawn. If we can use the modified Carnot expression, Eq. (5.5), the adiabatic head can be determined from work, since

$$-W = F_r h_a/778.3\eta_a \tag{5.14}$$

where F_r is the mass flow of refrigerant. Combining (5.5) and (5.14) eliminates $-W$, giving

$$F_r h_a/778.3\eta_a Q_c = (T_3 - T_1)/\eta_c T_1$$

Substituting latent heat of vaporization $H_1 - H_0$ for Q_c/F_r, adiabatic head can be found:

$$h_a = 778.3(\eta_a/\eta_c)(H_1 - H_0)(T_3 - T_1)/T_1 \tag{5.15}$$

If efficiency factors are reasonably constant, h_a is seen to vary linearly with $T_3 - T_1$.

Next, $T_3 - T_1$ varies with heat load, as indicated by subtracting Eqs. (5.6) and (5.7):

$$T_3 - T_1 = T_s - T_c + Q_c[1/U_c A_c + (1/U_s A_s)(1 - W/Q_c)] \tag{5.16}$$

The importance of maximizing the UA product in both heat exchangers then becomes apparent.

To illustrate the general shape of a refrigeration load line, Table 5.1 and Fig. 5.8 have been prepared, based on a cycle in which Refrigerant 11 is used to pump heat from chilled water at 50°F to air at 90°F. A 10-°F temperature

Table 5.1

Adiabatic Head and Volumetric Suction Flow versus Heat Load

Q_c (%)	T_3 (°F)	T_1 (°F)	$-W/Q_c$	h/η (ft)	V_1 (%)
100	140	40	0.27	12,600	100
50	113	45	0.18	9400	41
0	90	50	0.10	5500	0

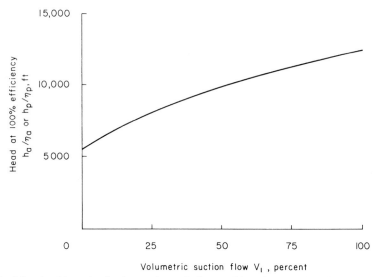

Fig. 5.8. A refrigeration load curve is nearly linear, with an intercept based on source–sink temperature difference.

drop is taken across the chiller at full load, and a 50-°F drop across the condenser. The head is given in terms of compressor efficiency—either adiabatic or polytropic head and efficiency may be used.

As anticipated, the load line is reasonably linear, elevated by the source–sink temperature difference. If this difference can be reduced, head will be reduced, thereby saving power and improving the turndown capability of the compressor. The slope of the load line varies with the heat-transfer areas and coefficients of the two exchangers. Increasing these parameters will lower the curve, also reducing power and improving turndown.

Recognize that the curve shown in Fig. 5.8 can be obtained only with the energy-saving control system of Fig. 5.3. As the heat load is reduced, suction pressure (and therefore evaporating temperature T_1) is allowed to rise, and condensing temperature T_3 is allowed to fall. If the uncoordinated system of Fig. 5.1 is used instead, suction pressure and condensing temperature are held constant, forcing the compressor head to remain constant at all loads. This not only uses more power at reduced loads, but raises the load level at which the surge limit is encountered, thereby reducing turndown. Figure 5.9 compares the two situations.

However, Fig. 5.9 shows only the effect of heat load on the two systems. Variations in source and sink temperatures can have a still more profound influence. If the ambient temperature were to fall, the entire curve for the coordinated system would fall, further reducing power and the surge limit.

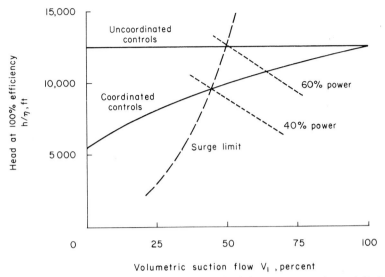

Fig. 5.9. An uncoordinated control system, by maintaining constant suction and discharge conditions, requires more power and decreases the operating range of the compressor.

The effect of raising the control point T_c would be similar but less dramatic, because it is unlikely to change as much as ambient temperature. By contrast, the uncoordinated controls of Fig. 5.1 protect the refrigeration unit against ambient changes, fixing the load line in a single position.

B. DEEP COOLING

While single-stage refrigeration units are by far the most common types encountered in industry, there are many applications they cannot satisfy. As the source–sink temperature difference increases, compression ratios must also increase. To improve compression efficiency, interstage cooling becomes necessary.

At some point, the use of a single refrigerant becomes inefficient. If compressor suction pressure is forced lower, equipment and piping must increase in size and therefore cost. At higher discharge pressures, costs also increase; furthermore, as criticality is approached, the latent heat of the refrigerant diminishes, requiring higher circulation rates for a given heat flow.

There are several alternatives to the single-refrigerant system, some of which will be presented to examine their control. Two different refrigerants may be used, separated by a heat-transfer surface, in what is known as a "cascade" system. Another approach is the use of a multicomponent mixture, which fractionates as it proceeds through the cycle. Refrigeration can also be obtained by expanding gases rather than liquids. Each method has its own problems and opportunities for energy conservation.

1. Cascade Systems

In a cascade system, a higher-boiling refrigerant is used to condense a lower-boiling one, as shown in Fig. 5.10. In this example ethylene is evaporated at perhaps $-150°F$, compressed, and condensed by boiling propane at about $-40°F$. The propane vapor is in turn compressed and condensed against cooling water at 80°F. A third stage can be added wherein methane is

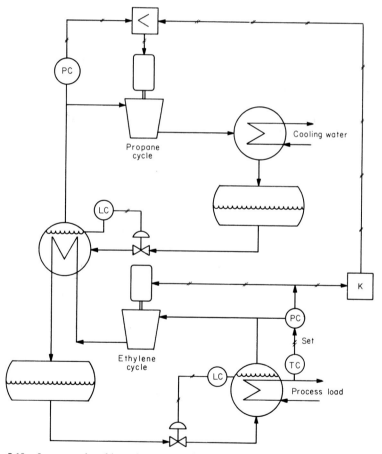

Fig. 5.10. In a cascade refrigeration system, the work load should be equitably distributed between stages.

condensed against the ethylene, allowing temperatures as low as $-250°F$ to be reached. A three-stage system such as this applied to the liquefaction of natural gas is described in Ref. (2).

Typically, each stage of a cascade system is controlled as if it were independent of the other stages. This can result in operation that is far from optimum. Although essentially the same heat load is passed serially from one stage to the next, the work required to pump that particular flow of heat may not be equitably distributed. If the temperature at the point of heat transfer between stages is too low, the high-temperature stage may be more heavily loaded than the low-temperature stage. The adiabatic efficiency of a compressor decreases as its compression ratio increases, per Eq. (4.12). Then the compressor with the higher ratio would require more power to transfer the same amount of heat than would be saved by the compressor with the lower ratio. An optimum interstage temperature exists where the load is divided in such a way that a minimum total power is needed to pump the required flow of heat. This will be near the point at which compression ratios are equal or temperature differences are equal, but not necessarily at it. The higher-temperature stage must pump more heat than the lower-temperature compressor, by the work that the latter introduces. In addition, the two fluids differ in their thermodynamic properties.

The control system in Fig. 5.10 offers a simple solution to this optimization problem. The temperature controller on the chilled process stream sets the low-temperature compressor speed through its suction-pressure controller. Then the other compressor speed is set in ratio to it, with the ratio adjusted to reach an observed optimum. What this system lacks in accuracy it gains in simplicity. Additionally, optimization systems ordinarily do not demand extreme accuracy because, in most cases, operating cost is relatively insensitive to the manipulated parameter in the optimum region.

Both compressors can benefit from favorable ambient conditions. As the high-stage discharge temperature falls, its uncontrolled suction pressure will follow. This in turn will cause the discharge temperature of the low-temperature stage to fall, decreasing its suction pressure. The suction pressure controller will then reduce the speed of *both* compressors proportionately.

Equitable distribution of load also improves the turndown of the entire system. The heat load at which a surge limit is encountered will be lowest if both compressors reach their limit together. Otherwise either one or the other will meet a limit at a higher load. In fact, the speed ratio could simply be set to make both compressors reach their surge limits together, and the system would probably operate quite close to the true optimum at all speeds.

In many cascade systems, the higher-temperature refrigerant serves other users as well as condensing the lower-temperature refrigerant. In this case, its maximum evaporation pressure would have to be selected to satisfy the most demanding user, as was done with the multiple side-load unit in Fig. 5.4.

2.   Mixed-Refrigerant Systems

The principal advantage of a mixed-refrigerant system is that deep cooling can be achieved with a single compressor. The compressed mixture is liquefied fractionally, in countercurrent heat exchange with liquid evaporating at lower pressure. The flowsheet as shown in Fig. 5.11 is a simplified description of a system designed for liquefaction of natural gas. The refriger-

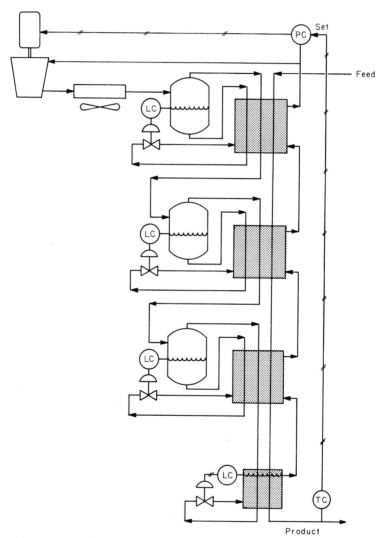

Fig. 5.11. In a mixed-refrigerant system, a temperature gradient is established by fractionation of the refrigerant at essentially constant pressure.

ant is a mixture of paraffin hydrocarbons from methane through butanes, with a small percentage of nitrogen.

After one (as shown) or two stages of compression and atmospheric cooling, the mixture fractionates into liquid and vapor. The liquid is cooled against vapor rising from colder sections, and then flashed to the vapor compartment of the same exchanger to provide additional chilling. The vapor stream from the separator is partially liquefied in the same exchanger, and sent to a second separator. The process is repeated at each exchanger—only four are shown here, but more may be used. The feed gas to be liquefied passes sequentially through the exchangers.

The pressure in the condenser tubes is essentially that of the compressor discharge, while the shells are at suction pressure. This pressure difference produces a temperature difference between tube and shell of each exchanger. However the temperature gradient from the warmest to the coldest exchanger is a function of the composition of the mixture.

In fact, composition control is the key to optimizing the performance of the system. If suction and discharge pressures are too high at a given load, heavy components should be added to the mixture or light components removed, depending on inventories elsewhere. Light components may be removed by venting the coldest sections where they concentrate.

A low concentration of an intermediate component such as propane or ethane will appear as insufficient liquid flowing through the expansion valve in that section. In a natural-gas liquefaction plant, the necessary components may be extracted from the feed gas itself.

The optimum refrigerant composition will change with the composition of the feed gas. Some adjustment is also required seasonally, as the plant load and ambient conditions change. The controls that operate the compressor and the atmospheric condenser are basically no different than for a single-stage system. To take advantage of colder ambient conditions and feed gas, some of the heavier components should be removed during the winter, and replaced during the summer. Otherwise compressor pressures might move beyond allowable limits when extremes of weather and load are encountered. The added dimension of composition adjustment makes the mixed-refrigerant system potentially more efficient than a cascade system, in that it can be adjusted optimally for each set of operating conditions. How closely the optimal is approached in practice depends on how rapidly and widely conditions change, and the difficulty and cost of adjusting refrigerant composition.

3.   Gas-Phase Refrigeration

Refrigeration can also be achieved by removing energy from a cooled, compressed gas in an expansion engine. This is the method used to liquefy air, with heat being rejected to a conventional ammonia cycle in cascade. It is also possible to achieve cooling by isenthalpic expansion of a gas across a valve, depending on its Joule–Thompson coefficient.

If an ideal gas is expanded through a restriction at constant enthalpy, its temperature will not change. The Joule–Thompson coefficient μ,

$$\mu \equiv (\partial T / \partial p)_H \tag{5.17}$$

is zero for an ideal gas. Most real gases have positive Joule–Thompson coefficients, while those of hydrogen and helium are negative. Some, like carbon dioxide, have negative coefficients at very low temperatures. Table 5.2 lists coefficients for some common gases at $-50°C$ and 60 atm.

Table 5.2

Joule–Thompson Coefficients for Common Gases at $-50°C$ and 60 atm

Gas	μ (°C/atm)	Gas	μ (°C/atm)
Air	0.32	Hydrogen	-0.018
Carbon dioxide	-0.015	Nitrogen	0.30
Helium	-0.061	Methane	0.71

As the table indicates, air, nitrogen, and methane can be used as refrigerants under conditions of isenthalpic expansion, while the others cannot. This is the basis of the Linde process used for air liquefaction and separation at the turn of the century.

The Claude process, using expansion through an engine, is much more efficient. A greater temperature drop is achieved, and work is recovered for compression as well. Figure 5.12 gives a simplified description of a process for liquefying carbon dioxide from a mixture of gases, using an expander.

The mixture is compressed and cooled against water and refrigerant. The final stage of cooling is regenerative—gas exhausted through an expander is reduced sufficiently in temperature to liquefy the product. A process of this sort would ordinarily be controlled at constant pressure by throttling the exhaust, and at constant temperature by throttling the refrigerant. In the absence of exhaust-gas throttling, liquefaction pressure would vary somewhat with load, as shown in Fig. 5.13.

If the load were constant, as if the exhaust gas were throttled through a fixed restrictor, discharge pressure and flow would follow that load line as compressor speed varied. An expander typically produces a linear relationship between inlet pressure and mass flow, as seen in the characteristics of a steam turbine coupled to a generator. When its speed is increased, the pressure drop decreases for a given flow, reflecting a decrease in internal resistance. Since the expander and compressor are connected to the same shaft, a

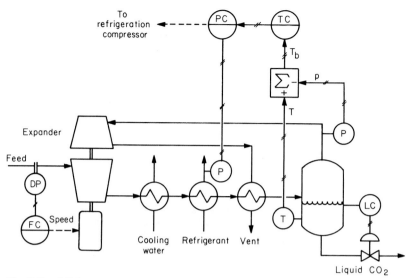

Fig. 5.12. Minimum energy will be used when the boiling point of the liquid product is controlled at the variable inlet pressure of the expander.

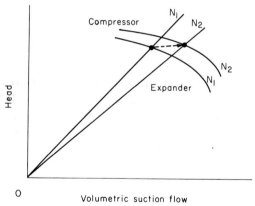

Fig. 5.13. A compressor–expander combination on the same shaft provides some regulation over pressure.

change in speed will cause both their curves to shift, in a manner that tends to regulate pressure. Pressure will still rise with an increase in speed, but not as much as it would with a constant load; the principal effect will be a change in flow.

Maximum efficiency will be achieved if the expander is not restricted by a throttling valve, which means the pressure must be allowed to vary. Then control over the liquefaction process must be exercised by adjusting the amount of supplementary refrigeration provided. In essence, the composition of the liquid product is to be controlled. This can be accomplished by relating its pressure and temperature along a contour of constant composition—essentially a vapor-pressure or dew-point curve. For moderate variations in temperature, the curve is nearly linear. The dew point can then be calculated as a function of observed temperature T and pressure p:

$$T_b = T - \frac{\partial T}{\partial p}\bigg|_x (p - p_b) \tag{5.18}$$

where T_b is the dew point of the mixture at base pressure p_b, and the partial derivative represents the slope of the curve at constant composition x.

C. STEAM REFRIGERATION

Refrigeration from steam at relatively low pressures is commonly used for air-conditioning moderately sized buildings and industrial facilities. The efficiency of these systems is not high. Basically, they pump heat from chilled water to cooling water much like a mechanical refrigeration unit, using the available work in steam rather than mechanical work. But much more low-pressure steam is required to generate a ton of cooling than high-pressure steam, simply because of the difference in available work. There is no advantage to providing part of a plant's cooling needs by a steam-driven compressor, with the balance produced by exhaust steam from the turbine. Refrigeration requires work, rather than heat, so there is no optimum balance between refrigeration loads, as exists between work and heat.

Refrigeration from low-pressure steam is therefore most commonly used for air-conditioning facilities that use steam for heating. Then the steam demand does not vary greatly from winter to summer. If electrically powered air-conditioners were used, the facility's electrical demand would be high in summer and its steam demand high in winter—an imbalanced condition.

There are essentially two types of steam refrigeration—absorption, and jet-ejector refrigeration. Both are based on vacuum evaporation of water.

1. Absorption Cycles

In an absorption system, a brine of lithium bromide or similar hygroscopic salt reduces the vapor pressure of the water in which it is dissolved. This promotes evaporation of distilled water, drawing its latent heat from the chilled water as shown in Fig. 5.14.

The unit is split into two sections: the upper, where the solution is concentrated, and the lower, where evaporation takes place. In the upper body, steam boils the brine at about 1.5 psia. The water distilled is condensed

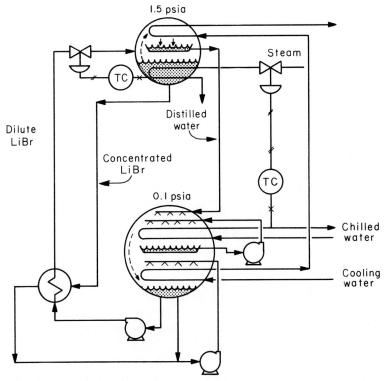

Fig. 5.14. Water vapor is absorbed into a brine at a low pressure in the lower body; the brine is reconcentrated in the upper body to complete the cycle.

against cooling water and flows to the evaporating section of the lower body. Concentrated brine exchanges heat with dilute brine flowing to the upper body, and strengthens the brine being used for absorption.

In the lower body, distilled water is recirculated at 0.1 psia, evaporating against the chilled-water coils. At that pressure, water boils at 35°F, producing chilled water in the 40–50°F range. The low absolute pressure is maintained by absorbing the water vapor into recirculated brine, with its latent heat transferred to a cooling-water coil. The pressure reached is therefore a function of the concentration of the brine and the temperature of the cooling water. The boiling point of the brine is typically controlled by admission of weak solution to the concentrator as shown.

The efficiency of converting work into heat in an absorption unit is relatively low. Reference (3) cites a flow rate of 18.7 lb/h of steam at 5 psig required to generate one ton of refrigeration at 40°F, using cooling water at 85°F. Under these conditions, the available work of the steam is 215 Btu/lb.

Then

$$\frac{-W}{Q} = \frac{18.7}{12000} \frac{\text{lb/ton-h}}{\text{Btu/ton-h}}(215 \quad \text{Btu/lb}) = 0.335$$

If the temperatures for the chilled water and cooling water are inserted in Eq. (5.5), along with the work–heat ratio above,

$$\eta_c = \frac{85 - 40}{460 + 40} \frac{1}{0.335} = 0.269$$

As with other refrigeration systems, the work–heat ratio can be improved markedly as the temperature difference between source and sink reduces. Reference (3) indicates that the steam flow required per ton of cooling decreases by 10% as the heat load falls from 100 to 40%. This improvement in performance is probably due to decreased temperature gradients across all heat-transfer surfaces. Consequently, the estimate of η_c above is probably about 10% low, in that T_s and T_c were used in its calculation, instead of the unmeasurable T_3 and T_1. Further reduction in load below 40% does not seem to offer any improvement in efficiency, probably due to deteriorating heat-transfer coefficients at lower velocities.

This data was all obtained at a 40-°F chilled water temperature. If this temperature control point is raised, the steam consumption per ton of refrigeration is reduced, and the capacity of the unit increased. With 85°F cooling and steam at 5 psig, an increase in T_c from 40 to 50°F increases unit capacity by 24% and reduces $-W/Q$ by about 4.5%, according to Ref. (3). According to Eq. (5.5), both should improve by 31%.

Performance is also improved as cooling water temperature falls. A drop from 85 to 75°F increases unit capacity by the same 24%, but reduces steam consumption only 2.2% (3). Again, Eq. (5.5) predicts a 28% reduction. Reference (4) however reports a reduction of 11% for a two-stage absorption unit for the same drop in cooling-water temperature.

The two-stage unit concentrates the absorbent by double-effect evaporation. Steam at 125 psig boils brine at about 5 psig. The vapor resulting from this stage of evaporation is then applied to the concentrator of a single-stage unit. The work–heat ratio is essentially identical to that of a single-stage unit, although steam usage is about 37% lower. In a plant where steam at 125–150 psig is more available than 5-psig steam, the additional stage is worthwhile.

If steam is accounted in the plant on an enthalpy basis alone, the two-stage unit will appear to be more economic. But if accounting is based on available work, there will be no difference between the two choices.

The controls as shown in Fig. 5.14 maintain chilled-water temperature by modulating steam flow. As steam flow increases, the temperature of the concentrated brine rises, causing the concentrator inlet valve to increase feed rate. This causes more concentrated brine to overflow into the absorber, raising the strength of the absorbent and thereby reducing the pressure in the evaporator. The subsequent reduction in boiling point increases the temperature gradient across the chiller tubes, thereby increasing heat transfer.

In most systems, cooling-water flow is fixed, as is the control point for chilled-water temperature. Improvement in efficiency is possible if appropriate adjustments are made, commensurate with the heat load and cooling-water temperature. Following the discussion of steam-jet refrigeration, optimization of these systems by cooling-water manipulation is presented. Then a program for managing chilled water systems is proposed, to maximize T_c as a function of load, and, with it, the efficiency of the unit.

2. Steam-Jet Refrigeration

Steam-jet refrigeration is less efficient than absorption refrigeration in the 40–50-°F range, but its efficiency improves markedly as the chilled-water control point is raised. Consequently, its most common service is in the range of 50–80°F, between absorption refrigeration and cooling-tower water.

Figure 5.15 shows the chilled water being cooled directly by evaporation in a vacuum chamber. Vapors are withdrawn through a steam-jet ejector

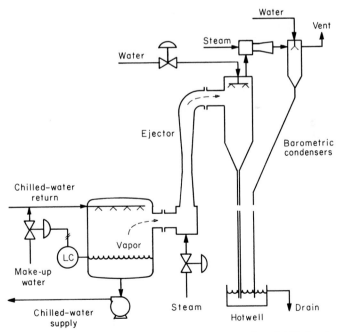

Fig. 5.15. Steam-jet ejectors provide vacuum evaporation of chilled water.

which discharges into a condenser. If the condensate is worth recovering, a surface condenser must be used, with a condensate pump. If not, a direct-contact barometric condenser may be used, discharging into a hotwell. The barometric leg must be 34 ft in length to keep air from backing into the vacuum chamber. Noncondensible gas such as infiltrated air is removed by a smaller single-stage or two-stage ejector, to maintain vacuum. The system may be operated either closed or open; i.e., chilled water may be returned or not. The open operating mode is less efficient in that make-up water would almost certainly be warmer than returning chilled water.

Evans (5) lists the utility requirements in terms of 85-°F cooling water and 100-psig steam per ton of cooling, as a function of chilled-water temperature. His data, converted to work–heat ratio, appear in Table 5.3. From this information, the steam-jet refrigerator seems to follow the Carnot equation much more closely than does the absorption unit.

Table 5.3

The Work–Heat Ratio for Steam-Jet Refrigeration Using 100-psig Steam and 85°F Cooling Water

Chilled-water temperature (°F)	Cooling water flow (gpm/ton)			
	4	6	8	10
40	—	—	0.647	0.584
50	0.663	0.492	0.449	0.426
60	0.418	0.352	0.327	—
70	0.286	—	—	—

For example, raising T_c from 50 to 70°F should reduce $-W/Q$ by a factor of

$$\frac{85 - 70}{460 + 70} \frac{460 + 50}{85 - 50} = 0.412$$

Dividing the W/Q ratios obtained at 4 gpm/ton and 50 and 70°F in Table 5.3 gives

$$0.286/0.663 = 0.431$$

Similar results are obtained at other temperatures and cooling-water rates.

This exercise indicates that the performance of the steam-jet refrigerator is predictable using Eq. (5.5), both with respect to source and sink temperatures. Being more sensitive to these variables than the absorption unit, its efficiency will be much more responsive to control adjustments. Considerable improvement in efficiency may then be realized under favorable conditions of load and coolant temperature. In this respect, it resembles a mechanical refrigeration unit.

The vapor pressure attainable in the evaporator varies with the compression ratio of the ejector and the backpressure into which it discharges. Neglecting the very small amount of infiltrated air likely to be encountered, the backpressure is essentially the vapor pressure of water at the condensing temperature. In essence, the ejector pumps heat from the evaporating temperature to the condensing temperature. Naturally, the condensing temperature is higher than the cooling-water temperature and is proportional to cooling-water flow. Hence increasing cooling-water flow will reduce the condensing temperature, and thereby improve the efficiency of the unit, as Table 5.3 indicates.

3. Optimization

Whenever a single controlled variable (such as chilled-water temperature) responds to two manipulated variables (such as steam and cooling-water flows), there is an opportunity for optimization. If the cooling water had no cost, its flow could be maximized, which would in turn minimize the amount of steam required to produce a given chilled-water temperature. Or if the steam were free, its flow could be maximized, which would allow cooling-water flow to be reduced to a minimum.

Actual plant situations lie between these extremes: neither utility is free. Hence there exists some combination of the two which will provide chilled water at the necessary temperature at a minimum operating cost. Neither the optimum cooling-water flow nor the optimum steam flow are likely to be constant, however. The refrigeration system is subject to variations in heat load, chilled-water temperature, and cooling-water temperature. As these parameters vary, *both* manipulated variables will require readjustment to maintain optimum conditions. An analysis of the data in Ref. (5) will reveal the coordination required between the two flow rates.

If steam flow in pounds per ton-hour is plotted against condenser cooling-water flow for each value of T_c, a series of hyperbolas is formed, as in Fig. 5.16. The general relationship for the curves is

$$F_s = a/F_w \tag{5.19}$$

where coefficient a varies with T_c.

The cost of operating the unit is the sum of steam and cooling water costs:

$$\$ = v_s F_s + v_w F_w \tag{5.20}$$

If Eq. (5.19) is a reasonable model of the process, a substitution can be made for F_s in terms of F_w:

$$\$ = v_s a/F_w + v_w F_w \tag{5.21}$$

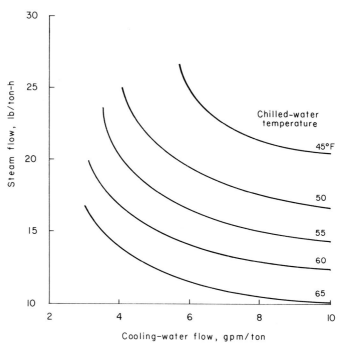

Fig. 5.16.   Steam usage per ton of refrigeration is affected by both cooling-water flow and chilled-water temperature.

Equation (5.21) indicates the presence of a minimum. When (5.21) is differentiated and set to zero, the optimum value of F_w may be found:

$$d\$/dF_w = -v_s(a/F_w^2) + v_w = 0, \qquad (F_w)_{opt} = \sqrt{av_s/v_w} \qquad (5.22)$$

The minimum point in a hyperbolic equation such as (5.21) happens to be coincident with equality of the individual terms. That is, when the cost of steam and cooling water are approximately equal, operation is optimum. To demonstrate, let

$$v_s F_s = v_w F_w, \qquad v_s(a/F_w) = v_w F_w, \qquad F_w = \sqrt{av_s/v_w}$$

If this method of determining the optimum water flow is valid, then controlling at the optimum can be achieved simply by setting water flow in the proper ratio to steam flow:

$$(F_w)_{opt} \simeq (v_s/v_w)F_s \qquad (5.23)$$

To validate this approximation, the optimum cooling-water flow was found using an arbitrary set of costs, applied to the data from Fig. 5.16. Figure 5.17 was generated therefrom, using costs of $0.10/1000 gal for water

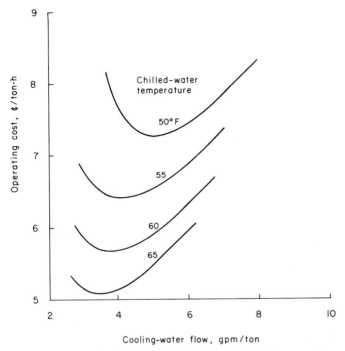

Fig. 5.17.   There is an optimum combination of steam and cooling water for each temperature.

and \$2.00/1000 lb of steam. The optimum cooling-water flow at 55°F chilled-water temperature occurs at about 4.6 gpm/ton. At that point, the cost of the cooling water is

$$4.6 \quad \text{gpm/ton} \times 60 \quad \text{min/h} \times 10¢/1000 \quad \text{gal} = 2.76¢/\text{ton-h}$$

From Fig. 5.16, the concurrent steam flow is about 18.6 lb/ton-h, costing

$$18.6 \times \$2.00/1000 \quad \text{lb} = 3.72¢/\text{ton-h}$$

The two costs are decidedly unequal at the optimum. If the control system is operated to make the two costs equal, it will drive the cooling-water flow to 5.55 gpm/ton and the steam flow to 16.9 lb/ton-h, raising the total cost from the optimum of 6.48¢/ton-h to 6.71¢/ton-h. The penalty for sub-optimum operation is an increase in cost of about 3.5%. This increment of saving is not likely to justify a more complex control system.

The control system required to equalize the cost of the two streams is shown in Fig. 5.18. Measured steam flow sets the cooling-water flow through a ratio flow controller (FFC). The flow ratio would be set to satisfy Eq.

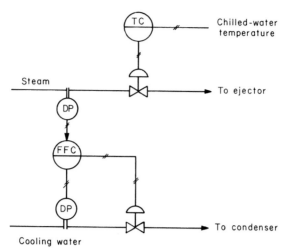

Fig. 5.18.   Controlling cooling-water flow in ratio to steam flow approaches optimum operation.

(5.23) if equal costs did in fact result in minimum cost. Since it does not, the ratio may be adjusted to a lower value, more in keeping with the costs determined by finding the actual optimum. While this ratio may not be exact for all chilled-water temperatures, it should approach the optimum more closely than the equal-cost ratio.

Variations in the chilled-water control point and in cooling-water temperature should have about the same effect on the system. The temperature controller will respond by readjusting both steam and water. Operating cost may change, but should still remain near optimum.

D.   BRINE AND CHILLED-WATER SYSTEMS

In plants where there are many refrigeration users, distribution of cooling is often carried out with brine or water, rather than by evaporating refrigerant at each user. This practice saves in the size of piping in that liquid rather than vapor is returned to the refrigeration unit. It also reduces the inventory of expensive refrigerant, and the risk entailed with its leakage.

Where subfreezing temperatures are encountered, the medium must be properly conditioned. Glycol–water mixtures are common in this service. Although certain organic fluids have been used, their cost and hazards cannot be justified, except for very low temperatures. Water is used for all systems not requiring temperatures below 35°F. Most central air-conditioning plants distribute their cooling as chilled water at 40°F.

One of the purposes of the following investigation is to determine the optimum temperature for brine or chilled water as load and ambient conditions vary.

1. Two-Way Valving

Chilled water or brine may be directed to users by means of either two-way or three-way valves. A two-way valve throttles the flow of the medium, stopping it altogether if no cooling is required. A three-way valve diverts part of the medium around the user such that its flow from the supply header to the return header tends to be relatively constant, regardless of the cooling load. The system with three-way valves is more efficient from the standpoint of the refrigeration unit. Maintaining a constant flow through the coils of an absorption unit or through the evaporator of a steam-jet unit will help keep its efficiency high at low heat loads. On the other hand, pumping costs are higher at low-load conditions. This may be an important consideration when transporting chilled water over long distances.

With either arrangement, the efficiency of the refrigeration unit will be improved if the temperature of the chilled water or brine is as high as the various users can accommodate. The needs of each user are best expressed as the position its control valve assumes in attempting to regulate the process using the cooling. If only a few users draw from a source of supply, their individual valve positions can be compared, with the highest controlled by adjusting the set point of the supply temperature, as shown in Fig. 5.19. The figure shows both two- and three-way valves being employed, for purposes of illustration—they are not normally combined.

Two-way valving does present a control problem: as flow past the temperature sensor decreases, the dead time in its control loop increases, destabilizing the loop. A bypass is often added to systems with two-way valves to maintain a minimum flow through the chiller and past the temperature sensor.

In situations where many similar users are drawing from a common supply, the problem of selecting the highest of a multitude of valve positions may be circumvented. If all the valves are of the two-way type, flow will vary directly with heat load, under conditions of constant supply temperature. A valve-position control system would attempt to maximize valve positions, and therefore flow, by raising supply temperature to its highest acceptable level. The same objective can be reached by a flow controller setting supply temperature. As the users reduce their individual flow rates due to a general decrease in cooling load, the flow controller will respond by gradually raising supply temperature. In the steady state, total flow will be restored, with the decrease in cooling load being absorbed by a temperature change, rather than the customary flow change. Like a valve-position

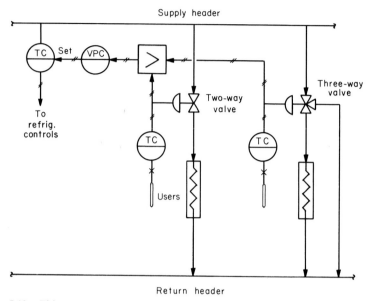

Fig. 5.19. Either two-way or three-way valves are used to distribute chilled water or brine; they are not normally mixed in the same plant.

controller, this flow controller requires integral action alone, with a time constant of several minutes—long enough for the users *and* the refrigeration unit to respond.

There is no guarantee that controlling total flow satisfy all users—it is actually only capable of satisfying the average user. Some measure of protection may be provided by reducing the flow set point somewhat below the maximum attainable. This should be acceptable for most air-conditioning systems, where cooling loads are generally evenly distributed, being affected similarly by ambient conditions and occupancy. In cases where one or two critical users may occasionally demand substantially more cooling than the average, a valve-position controller can be applied to them alone, as shown in Fig. 5.20. Its output could then override that of the flow controller to insure that the selected user is satisfied. During override conditions, total flow would fall below its set point as the other users reduce their demand for refrigeration.

2. Three-Way Valving

Users equipped with three-way valves must be treated in a different manner, because total flow is not affected by cooling load. In this situation the total cooling load must be inferred from a measurement of the tempera-

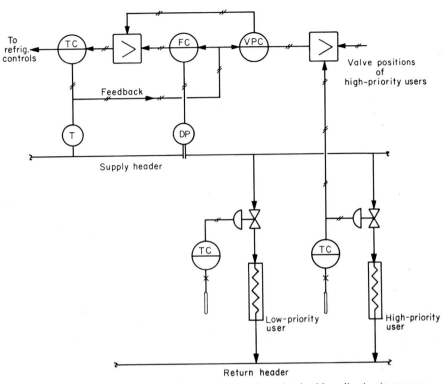

Fig. 5.20. A constant flow of cooling medium can be maintained by adjusting its temperature; high-priority users may override the flow controller.

ture T_{cr} of the returned coolant:

$$\sum Q_c = F_c C(T_{cr} - T_c) \qquad (5.24)$$

where F_c is the flow of chilled water or brine through the supply header, and C is its heat capacity.

The average user will transfer an average heat flow $\bar{Q}_c$ in proportion to the difference between the coolant temperature and the user-controlled temperature T_n:

$$\bar{Q}_c = K(T_n - T_c) \qquad (5.25)$$

This statement assumes a constant flow of coolant past the heat-transfer surface, which is not normally justifiable. However, the purpose of the control system is to maintain a constant flow through the average user.

The total heat-transfer rate is simply the number of users multiplied by the average rate, provided that all users have the same control point:

$$\sum Q_c = n\bar{Q}_c = nK(T_n - T_c) \qquad (5.26)$$

Combining (5.24) and (5.26) gives T_c as a function of the temperature difference between coolant supply and return:

$$T_c^* = T_n^* - G(T_{cr} - T_c) \tag{5.27}$$

with G replacing the group F_cC/nK. Again, the asterisks denote set points rather than measurements. The program appears graphically in Fig. 5.21.

The temperatures appearing in Fig. 5.21 were chosen to illustrate a typical air-conditioning system whose control point is set at 80°F. Coolant at 50°F is required to remove the full-load heat flow, causing a rise in coolant temperature of 15°F. Slope G is therefore 2.0.

Consider an air-conditioning system undergoing a change from full to $\frac{2}{3}$ cooling load. Controlled temperature T_n begins to fall, causing more coolant to be diverted around its heat exchanger. This will lower T_{cr}, reducing $T_{cr} - T_c$ from 15 to 10°F to satisfy the new heat balance. Without the program of Fig. 5.21, the system would then come to rest.

With the program, the system responds by raising T_c^* from 50 to 60°F. This change is gradual, because it requires the entire refrigeration unit to respond. As T_c slowly rises, T_n will tend to rise, and its temperature controller will then reduce the percentage of coolant diverted. Eventually the system will come to rest with the same diversion as at full load and with T_c at 60°F rather than 50°F; $T_{cr} - T_c$ will hold at 10°F, *raising* T_{cr} to 70°F. Without the program, it would have *fallen* to 60°F. This is bound to have a substantial impact on the efficiency of the refrigeration unit.

The program can be implemented as described in Fig. 5.22. There are two adjustments: T_n^* and G. If all user control points are not identical,

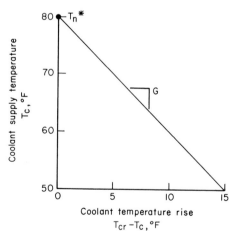

Fig. 5.21. With a system of three-way valves, coolant temperature should be set in relation to its rise.

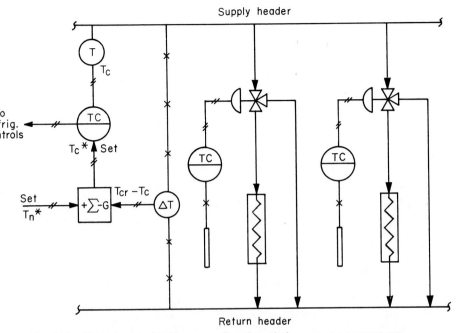

Fig. 5.22. This system maintains average flow through the users at a constant rate.

T_n^* must be set to favor the lowest—otherwise the lowest-temperature user could not be controlled at low load. The slope G should be adjusted to hold the average position of the user valves nearly full open. In replacing the functional group $F_c C/nK$, the setting of G determines the ultimate value of nK, since F_c and C are essentially constant. If the number of users in service, n, does not change, then G determines the average heat-transfer capability K of the average user, which is a function of flow through the user. Consequently G determines the average valve position. If it is set too low, some valves will be fully open, and temperature control at those points will be sacrificed. If it is set too high, T_c will be colder than necessary, and all valves will be diverting excessively.

As with the system of Fig. 5.20, overrides may be added to the supply temperature set point, based on the valve positions of high-priority users.

3. Variable-Rate Accounting

While the control systems described above enhance the efficiency of refrigeration facilities, they present a problem to those who sell energy. Consider a central facility providing steam and chilled water to many users such as stores, shops, apartments, etc. Traditionally, these services are controlled at specified conditions of pressure and temperature. Billing is applied on an energy basis. There may be no explicit credit given for the return of steam condensate, assuming that its heat content is constant. The same could be true of chilled water if it is returned at a constant temperature. Then billing may be based on flowrate information alone, because the energy transferred per unit flow is essentially uniform.

If the temperature of the return water is not constant, which is more often the case, the energy flow must be calculated as the product of flow and temperature difference between supply and return water, following Eq. (5.24). This is a rate equation: Q_c is expressed in heat flow, e.g., Btu's per hour. The heat flow signal must be integrated with respect to time, yielding a totalized quantity of energy, e.g., Btu, displayed or printed by a counter.

If chilled water is now supplied at a variable temperature to conserve energy, then heat-flow billing no longer reflects the true cost of refrigeration. If billing is maintained on a heat-flow basis, the savings in refrigeration power when 55-°F water is supplied, for example, would not be passed on to the user. The user should be compensated for variations in supply temperature, consistent with the cost of generating the cooling.

Because all refrigeration systems transfer heat by the application of work, billing should be based on power rather than heat flow. The power required to transfer a given flow of heat may be estimated using source and sink temperatures:

$$-W/Q_c = (T_s - T_c)/\eta T_c \qquad (5.28)$$

where T_s might be ambient or cooling-water temperature, as appropriate. Efficiency factor η would be determined by observing W and Q under a given set of operating conditions. This expression should be an acceptable model for mechanical and steam-jet refrigeration units, although absorption units may require a modification to achieve a better fit.

Then power can be calculated from measured heat flow:

$$-W = Q_c(T_s - T_c)/\eta T_c, \qquad -W = (FC/\eta T_c)(T_s - T_c)(T_{cr} - T_c) \quad (5.29)$$

The billing calculation is shown implemented with analog instrumentation in Fig. 5.23. A digital system would be more accurate but would not function continuously, calculating power at selected time intervals of 15 min or longer.

A cost comparison of variable-rate billing against constant-rate billing should reflect the savings attributable to variable-temperature operation. Using the example given in Fig. 5.21, the power requirements for cooling can be evaluated at full, $\frac{2}{3}$, and $\frac{1}{3}$ loads. The results are presented in Table 5.4, for a cooling-water temperature of 80°F.

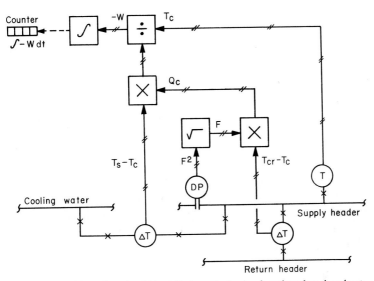

Fig. 5.23. Refrigeration should be billed on the basis of work rather than heat.

The constant-temperature system would show a power consumption directly proportional to the fractional heat load, as might be expected. However the variable-temperature system has a power consumption proportional to the *square* of the fractional heat load, as indicated in Table 5.4. It is therefore less costly for all operating conditions except full load, with the difference increasing as the load is reduced.

If the probability of operating at any given load is the same as that of operating at any other load, the average savings achievable with variable-temperature control may be determined. Let the operating cost for constant-temperature conditions be kQ_c, and that for variable-temperature conditions be kQ_c^2. Integrating the cost difference $\Delta\$$ uniformly over the entire range

Table 5.4

Power Requirements for Cooling a Variable-Temperature System

Heat load (%)	Supply temperature (°F)	Return temperature (°F)	$-W/Q$ (% of max.)	Power (%)
100	50	65	100	100
67	60	70	65	43
33	70	75	32	11

of Q from 0 to 100% gives

$$\Delta\$ = \int_0^1 (kQ - kQ^2)\,dQ = k\left(\frac{1}{2} - \frac{1}{3}\right) = \frac{k}{6}$$

Dividing by the integrated operating cost $ of the constant-temperature system gives

$$\frac{\Delta\$}{\$} = \frac{k/6}{k/2} = 33\%$$

So if the heat load for the facility were completely random in distribution, variable-temperature operation could be expected to save about 33% of the energy used with constant-temperature control.

Needless to say, the savings are realized where the cooling is generated, although they are brought about by user controls. Without a variable-rate accounting, no incentive for savings exists.

REFERENCES

1. Shinskey, F. G., "Process-Control Systems," pp. 221–222. McGraw-Hill, New York, 1967.
2. Harper, E. A., J. R. Rust, and L. E. Dean, Trouble-free L. N. G., *Chem. Eng. Progr.* (Nov. 1975).
3. "Absorption Liquid Chillers, Model EL," Technical Manual. York Corp., York, Pennsylvania, 1966.
4. Farwell, W., More cooling at less cost, *Plant Eng.* (Aug. 22, 1974).
5. Evans, F. L. Jr., "Equipment Design Handbook for Refineries and Chemical Plants," Vol 1, pp. 170–171. Gulf Publ., Houston, 1971.

Part

III

MASS-TRANSFER OPERATIONS

Chapter

6

Evaporation

Much of the energy consumed by industry serves to separate and refine products through mass-transfer operations. These operations involve changing the composition of mixtures, principally through diffusional rather than mechanical processes. Hence subjects like filtration and classification are not generally included under this heading.

Because the primary thrust of this book is the conservation of energy, the most energy-intensive mass-transfer operations are selected for study. In all these operations, the principal goal is the establishment of more-ordered products from a less-ordered feedstock. The force customarily applied to bring about this reduction in entropy has been heat. Since the products themselves contain no more energy than the feed, all the heat applied to separate them is ultimately rejected to the environment. Consequently, there is a considerable opportunity to save energy by operating these processes in a more efficient manner. The method used to measure the efficiency of a separation process is an entropy analysis. Mass-transfer operations are not traditionally evaluated in these terms, yet it will be found to cast a new light on the subject and point the way to increased energy savings.

Evaporation has been chosen to introduce the subject of controlling mass-transfer operations because the process is easy to describe mathematically, and yet can be difficult to control. Substantial energy savings are possible using such techniques as vapor recompression. However they involve thermodynamic relationships which must be understood before controls can be effectively applied.

A. EVAPORATING SYSTEMS

Evaporation is the separation of a volatile liquid from a nonvolatile solid by the application of heat. It has been practiced for centuries to obtain salt from seawater tidal pools, using solar energy. In the New England "sugar house," maple sap is concentrated to syrup by boiling over a wood fire.

In most evaporators, the solid materials are more valuable than the solvent that is driven off. But evaporation is also used to recover fresh water from seawater and from industrial brines and sludges, whose solids contents have little value.

Evaporation is distinguished from drying as being conducted in the absence of air or other noncondensible gas. The equilibrium relationships in drying are complicated by this additional dimension.

1. Multiple-Effect Evaporation

Using a wood fire to make maple syrup is extremely inefficient. Even if heat losses were minimized and the fuel–air ratio carefully controlled, the process would still be inefficient. Modern industrial practice applies multiple-effect evaporation to concentrate sugars and syrups, wherein the energy introduced is used more than once.

A triple-effect evaporator for concentrating a dilute brine is shown in Fig. 6.1. If the product does not present handling problems due to a high

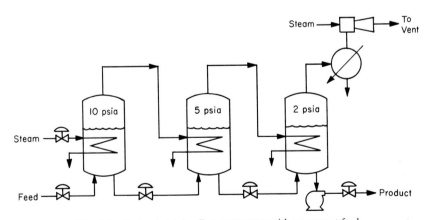

Fig. 6.1.   A simple triple-effect evaporator with co-current feed.

viscosity or suspended solids, and if it is thermally stable, large-capacity evaporator bodies may be used. They typically contain a large bundle of heat-transfer tubes known as a "calandria." A steam chest surrounds the tubes, and boiling liquid circulates through them in natural convection.

Vapor removed from the first effect is condensed by boiling liquid at a lower pressure in the second effect and so on. Less vapor is boiled than is condensed in any given effect because the latent heat of vaporization increases as pressure and temperature are reduced. In order to transfer the requisite flow of heat, a proportional temperature gradient must exist across each heat-transfer surface. This difference ultimately determines the pressures in the various effects, and the heat economy of the unit.

Products that are not thermally sensitive, such as seawater, may be evaporated using as many as 13 effects. Each added effect is slightly less efficient than the last, due to increasing latent heats and heat losses, making each increment of capital investment less attractive. Products that are temperature sensitive, such as food products, are limited to evaporation in two or three effects. The temperature of the last effect is limited by both the available cooling water and the viscosity of the product, whereas temperature at the first effect is limited by the thermal sensitivity of the solids. While dilute solutions may be evaporated in three or four effects, final products, leaving at 75–85% concentration, are typically processed in double-effect evaporators.

The viscosity problem can be mitigated by changing the direction of flow. Figure 6.2 shows both a countercurrent (reverse) flow evaporator and

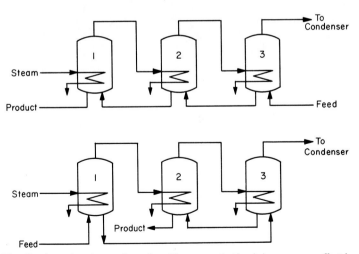

Fig. 6.2. Product viscosity can be reduced by concentrating it in a warmer effect in reverse flow (upper) or mixed flow (lower).

another with mixed flow. Note that the effects are always numbered from the highest to the lowest pressure.

There are a variety of other flow patterns in use. Black liquor from a pulp mill is concentrated in six effects of reverse flow, with an intermediate decanter for removing tallow. In some processes, two bodies will operate in parallel on the steam side and in series on the product side. This arrangement is seen in processing distillery wastes: effects 1A and 1B receive steam from the same header and discharge vapor to effect 2, yet syrup proceeds in reverse flow from 2 to 1B and then 1A.

Special designs are available for heat-sensitive materials to minimize the residence time of the product at elevated temperatures. In the evaporator shown in Fig. 6.3, heat transfer takes place through small-diameter tubing at high velocities. The heated solution then flows into a flash chamber where vapor and liquid are separated. Residence time is minimized by low liquid inventory and a once-through flow pattern, i.e., without circulation. These evaporators tend to be fast to start up but are sensitive to upsets and difficult to control.

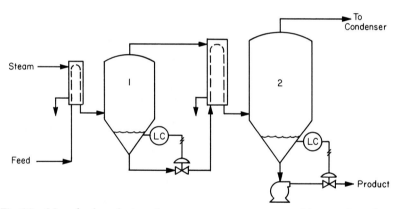

Fig. 6.3. Many food products such as corn syrup are concentrated in once-through evaporators with minimum liquid inventory.

2.   Material and Energy Balances

Consider an evaporator concentrating a mass flow of feed F_0 from an initial weight fraction of solids w_0 to a final of w_n. If the flow of final product is F_n, a material balance on the solids is

$$F_0 w_0 = F_n w_n \tag{6.1}$$

An overall material balance would include all of the vapor $\sum V_i$ removed from the various effects:

$$F_0 = \sum V_i + F_n \tag{6.2}$$

When these two expressions are combined, they yield the flow of vapor which must be driven off in order to reach the desired final concentration:

$$\sum V_i = F_0(1 - w_0/w_n) \tag{6.3}$$

The flow of vapor removed in the first effect is related to the steam flow to the first effect, and its thermal efficiency E:

$$V_1 \Delta H_1 = E V_0 \Delta H_0 \tag{6.4}$$

Similarly, the vapor removed from the second effect is related to first-effect vapor rate:

$$V_2 \Delta H_2 = E V_1 \Delta H_1 = E^2 V_0 \Delta H_0 \tag{6.5}$$

The sequence can be continued to n effects, and then the individual vapor flows summed:

$$\sum_{i=1}^{i=n} V_i = V_0 \Delta H_0 \left(\frac{E}{\Delta H_1} + \frac{E^2}{\Delta H_2} + \cdots + \frac{E^n}{\Delta H_n} \right) \tag{6.6}$$

The enthalpy ΔH_i gained by the vapor boiling in a given effect is not exactly equal to its latent heat of vaporization. In cocurrent flow, the solution coming from an upstream effect will be above its boiling point due to the reduction in pressure. However the vapor driven from each effect is superheated by the boiling-point elevation of the solution. From the standpoint of an overall energy balance, these factors are not great. The product from a co-current evaporator tends to leave at roughly the same temperature as the feed.

If latent heats are used for each ΔH in Eq. (6.6), the vapor–steam ratio can be readily calculated for any evaporator. Table 6.1 lists that ratio for evaporators up to eight effects, beginning with 125°F in the last effect, and assuming a temperature difference of 33°F per effect and an efficiency of 97%.

Table 6.1

Vapor-Steam Ratios versus Number of Effects of 97% Efficiency

Effect i	Temperature (°F)	Pressure (psia)	ΔH_i (Btu/lb)	$\dfrac{\sum V_i}{V_0}$	Number of effects n
n	125	2.0	1023	0.95	1
n-1	158	4.5	1004	1.85	2
n-2	191	9.5	983	2.71	3
n-3	223	18	963	3.52	4
n-4	256	33	941	4.28	5
n-5	289	57	918	4.98	6
n-6	322	92	893	5.62	7
n-7	355	143	866	6.21	8
n-8	388	215	837		

3. Boiling-Point Elevation

From the above discussion, it can be seen that the number of effects can be increased if the temperature drop per effect is reduced, allowing a greater steam economy. The temperature drop can be reduced to some extent by increasing the ratio of heat-transfer area to heat flow. But even with a large area or small flow (as at low-load conditions) there will remain a residual temperature drop due to boiling-point elevation.

The elevation in boiling point caused by a nonvolatile solid in solution follows the laws of physics so closely that it is used to determine the molecular weights of solid materials. For example, the boiling point of 1 kg of water at atmospheric pressure is elevated about 0.5°C by 1 gmol of a nonvolatile solute. This factor is not constant but changes with temperature:

$$\Delta T_b = (RT^2/\Delta H_v)x \tag{6.7}$$

where ΔT_b is the boiling-point elevation, R the universal gas constant, 1.987, T the normal boiling point in absolute units, ΔH_v the heat of vaporization, and x the mole fraction solute. (Units may be either English or metric.) Table 6.2 lists the boiling-point elevation for sucrose solutions at atmospheric pressure as given in the "Cane Sugar Handbook" (1).

Because boiling-point elevation is a function of mole fraction rather than weight fraction, solutes with lower molecular weights could be expected to have more influence. A 50 wt % solution of sucrose (molecular weight 342) is only 5 mol %, while a 50 wt % solution of caustic (molecular weight 40) is 21 mol %. Consequently, the caustic solution could be expected to exhibit a boiling-point elevation six times as high, or about 20°F at atmospheric pressure.

From the consideration of minimizing the temperature drop between effects, boiling-point elevation becomes important only in concentrated

Table 6.2

Boiling-Point Elevation of Sucrose Solutions at Atmospheric Pressure

Solids concentration w (wt %)	Boiling-point elevation ΔT (°F)	Solids concentration w (wt %)	Boiling-point elevation ΔT (°F)
0	0	75	12.38
33	1.33	80	16.99
50	3.22	85	23.93
60	5.36	90	34.69
70	9.31		

solutions. For example, a sugar solution might be evaporated from an initial 20 wt % to a final 80 wt % in three effects. Using the material and energy balance equations already presented, solution concentrations in the individual effects were calculated to be 27, 41, and 80 wt %, respectively. Boiling-point elevation in the three effects are then 1.1, 1.8, and 12.2°F, respectively, when corrected for temperature and heat of vaporization per Eq. (6.7). Consequently only the last effect has a boiling-point elevation which makes a significant contribution to the temperature difference between effects.

The pressure in any effect of an evaporator is determined by the temperature at which the vapor condenses. However the boiling point will be elevated above the condensing temperature of the vapor as a function of concentration, as demonstrated. Vapor boiled from a solution is then superheated by the boiling-point elevation. Thus in a cocurrent evaporator, the largest boiling-point elevation appears to exist between the temperature of the last effect and the cooling water. In a countercurrent system, the largest boiling-point elevation will develop between the first and second effects.

4. An Entropy Analysis

A triple-effect evaporator is not really as efficient as it might appear. A comparison of the entropy reduction in the feedstock against the entropy gained by the surroundings will bear this out. Consider a triple-effect sucrose evaporator concentrating feed from 20 to 80 wt % (Fig. 6.4). Let the steam conditions be represented by the first three effects in Table 5.1.

A material and energy balance on the process indicates that 3.6 lb of feed can be concentrated per pound of steam. Assuming that the feed and product

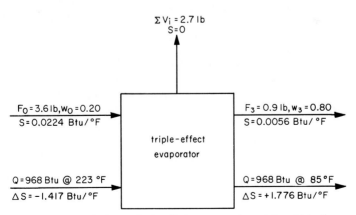

Fig. 6.4. An entropy analysis reveals that 21 times as much work is put into the evaporator as is obtained from it.

temperatures are the same, the entropy reduction achieved is related solely to composition. At 20 wt % solids, the feed contains 0.72 lb sucrose and 2.89 lb water per pound of steam. When divided by the molecular weights of sucrose (342) and water (18), the feed can be represented as consisting of 0.00211 mol sucrose and 0.16 mol water for a total of 0.162 mol. The mole fraction of sucrose is then 0.013. The entropy of the feed mixture using Eq. (1.21) is 0.138 Btu/°F per mole of feed, or 0.0224 Btu/°F per pound of steam. Applying the same evaluation to the product, consisting of 0.9 lb or 0.0121 mol at 0.174 mole fraction sucrose, yields an entropy of 0.462 Btu/°F per mole or 0.0056 Btu/°F per pound of steam. Since the entropy of the condensed vapor is essentially zero, there is a net decrease in entropy on the process side of 0.0224 − 0.0056 = 0.0168 Btu/°F per pound of steam.

Each pound of steam gives up its latent heat of 968 Btu in the first effect. That heat is ultimately rejected to the environment, as heat leakage and to condenser water. The entropy of the steam is reduced in condensing by 968 Btu/lb divided by its absolute temperature 460 + 223°F, giving a reduction of 1.417 Btu/lb-°F. The same amount of heat rejected to an environment at 85°F increases the entropy of the environment by 1.776 Btu/°F per pound of steam. The net increase in entropy of the heating and cooling system is 0.359 Btu/°F per pound of steam, 21 times the decrease in process entropy.

If all heat transfer surfaces were increased to minimize temperature gradients, the boiling-point elevation would still remain. Under these conditions, the total boiling-point elevation would be about 15°F. Then steam at only 100°F could perform the evaporation. Even at this minimum temperature difference, the entropy increase in the heating and cooling system is still 0.047 Btu/°F for the same quantity of feed, nearly 3 times the entropy reduction of the process.

5.   Vapor-Compression Cycles

An evaporator may be driven by work as well as heat—both mechanical and steam-jet compressors have been used in vapor-compression cycles. An example is the single-effect mechanical compression system described in Ref. (2). Shown schematically in Fig. 6.5, it is used to recover water from industrial effluents, leaving a slurry of salts suitable for landfill.

After being preheated by the distilled water, the feed is deaerated to remove air and carbon dioxide. It then joins the recirculated slurry flowing downward through a tube bundle, where heat is absorbed by condensing steam. Evaporation takes place at atmospheric pressure in that the system is vented through the deaerator.

Vapor leaves the slurry superheated by the boiling-point elevation of perhaps 4°F. The vapor is compressed to about 17.5 psia and 260°F, con-

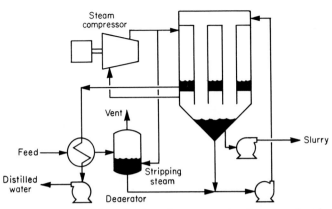

Fig. 6.5. Steam evaporated from the slurry is compressed and condensed against it in this brine concentrator.

densing at about 222°F, to provide a 6-°F temperature difference across the heat transfer surface. The condensed water then preheats the feed.

The method for estimating the work required by the compressor is described in relation to heat pumps in Chapter 5. Figure 5.2 and Eqs. (5.1)–(5.5) describe a refrigerant flowing in a closed cycle. Here, steam flows in an open cycle, but the same equations apply. The temperature difference between source and sink in the evaporator is simply the boiling-point elevation and the differential across the single heat-transfer surface. Consequently, a much lower work–heat ratio might be expected than is common to refrigeration systems.

In the vapor-compression cycle just described, an estimated 21.4 Btu of work is required to evaporate 1 lb of water at 216°F. The work–heat ratio for this process is then 0.022, which agrees exactly with the designers' estimate of 52 kWh of electrical energy per 1000 gal. When all auxiliaries are included—pump motors, stripping steam to the deaerator, and heat losses—the actual power consumption comes to 72 kWh/1000 gal, or a work–heat ratio of 0.030.

Considering that the electric power used to drive the plant is generated from fuel at a thermodynamic conversion of perhaps 32%, the thermal economy of the system is about 10.7 Btu of water distilled per Btu of fuel. This is equivalent to more vapor than can be removed per pound of steam in 15 effects of conventional evaporation. And since the point of diminishing returns is usually reached long before 15 effects, it would be safe to say that the single-effect vapor-compression evaporator is more efficient than any number of conventional effects. Naturally, boiling-point rise does influence the economy—the 4-°F rise in this example is lower than encountered in most other systems.

The net input of energy to the compressor ultimately leaves the evaporator as a difference in sensible heat between the products and the feed. Without another heat source or sink, the temperatures of the products will then float on the temperature of the feed. The compressor suction is held essentially at atmospheric pressure because the deaerator is vented. If stripping is inadequate, noncondensible gas will interfere with heat transfer and reduce the efficiency of the cycle. Excessive stripping will increase steam losses out the vent, and also reduce efficiency. For best performance, stripping steam should be manipulated to control differential vapor pressure in the deaerator, as measured by the DVP transmitter described in Figs. 3.6 and 3.7.

In certain syrup evaporators, minimum temperatures must be maintained to promote flow. Then additional steam may have to be injected into the cycle to control product temperature or absolute pressure in that effect.

Steam-jet ejectors are being increasingly used in the first effect of multiple-effect evaporators. Reference (3) describes such a system, where 150-psig steam rather than low-pressure steam is supplied to the first effect, as shown in Fig. 6.6. The ejector compresses some of the first-effect vapors to the pressure of the first-effect steam chest, as was done with the mechanical compressor. To generate 1 lb of steam at 14.7 psia requires 0.474 lb of 165-psia steam compressing 0.526 lb of first-effect vapor saturated at 180°F. A temperature difference of 32°F exists across the heat-transfer surface.

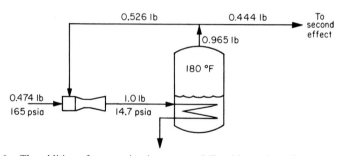

Fig. 6.6.   The addition of a steam-jet ejector essentially adds another effect to the system.

The pressure–enthalpy diagram of Fig. 6.7 helps to explain what happens in a jet compressor. When the two streams are combined, a mass and enthalpy balance must be reached. Thus, regardless of the thermodynamic efficiency of the compressor, the enthalpy of the mixture is determinable. If the mixing is carried out irreversibly, there will be no compression, and the mixture will simply be superheated to point A in the figure. This is of no advantage since the latent heat of the mixture will not be recoverable.

If the compression were completely reversible, entropy would be conserved. An entropy balance on the mixture indicates that it would then lie on the saturation curve at point B, compressed to about 31 psia. The actual mix-

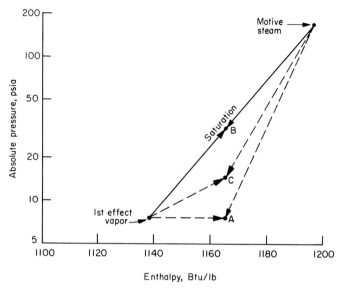

Fig. 6.7. Simple mixing would conserve enthalpy at the lower of the two pressures (A); reversible compression conserves both enthalpy and entropy (B); jet compression arrives at an intermediate point (C).

ture, however, lies between the two extremes, at point C. Entropy increases somewhat, and the mixture becomes superheated to about 243°F.

The jet compressor adds essentially one effect to the system, as the material balance in Fig. 6.6 indicates. Essentially double-effect evaporation is achieved in the first effect, in that 0.965 lb of vapor is driven off using only 0.474 lb of motive steam. However only 0.444 lb of vapor are delivered to the second effect. Then the second and all higher effects produce virtually the same amount of vapor per pound of motive steam that they would have without the compressor. In essence, the compressor is used in place of one effect of evaporation.

In actual practice, the work economy of this system is inferior to that of a simple multiple-effect evaporator. A triple-effect evaporator with a jet compressor will remove about 4.7 lb of vapor per pound of 165-psia steam. A seven-effect evaporator without compression can remove 5.6 lb of vapor per pound of 165-psia steam, according to Table 6.1. However temperature limits on products, capital costs, and availability of steam must all be considered in determining the optimum configuration for any process.

B. PRODUCT-QUALITY CONTROLS

Precise control over product quality is an integral part of any energy-conservation program. Products that are to be sold, or even those simply being sent to another part of the same plant, must meet certain established specifications. Failure to meet specifications generally entails some penalty— the product must be reprocessed, diverted to some alternate and less valuable service, or blended with higher-quality material. All of these choices are wasteful of energy. Reprocessing also reduces production capacity and creates operating problems in that a plant designed for a certain feedstock cannot efficiently accommodate off-specification product as feed. When a product must be diverted to a lower-grade service, much of the energy put into its preparation is lost, because less would be required to make the lower-grade product in a controlled situation. The loss in available work caused by blending was presented in Chapter 1, and need not be repeated here.

In order to avoid the penalties associated with off-specification product, most operators attempt to control with a very safe margin. The wider the margin allowed, the less likely will there be a violation of a quality limit, in the face of expected upsets in operating conditions. The width of the margin used is directly related to the variability in product quality. If a variation of $\pm\chi\%$ is commonly experienced, the operator will typically set his control point $\chi\%$ or more above specifications.

Several debits are associated with this mode of operation. Most evaporators are intended to yield a concentrated product having a certain percent solids. Exceeding specifications gives away more solids than contracted for, and uses more energy to achieve the higher concentration. Other penalties may also arise, such as product degradation and fouling of heat-transfer surfaces caused by overconcentration.

1.   Sensitivity to Disturbances

The control problem can best be appreciated by evaluating the sensitivity of product quality to variations in heat input and feed rate and composition. The material and energy balances already derived provide all the information necessary for this evaluation. Equation (6.3) may be solved for product quality:

$$w_n = w_0/(1 - \sum V_i/F_0) \tag{6.8}$$

Differentiating with respect to $\sum V_i/F_0$ will indicate the sensitivity of product concentration to the steam–feed ratio:

$$dw_n/d(\sum V_i/F_0) = w_0/(1 - \sum V_i/F_0)^2 = w_n^2/w_0 \tag{6.9}$$

Placing the differential in terms of concentrations facilitates comparison of sensitivities for various operating conditions. An additional help is to convert the denominator into a fractional steam–feed ratio, in which case the estimate is independent of the number of effects used:

$$\frac{dw_n}{d(\sum V_i/F_0)/(\sum V_i/F_0)} = \frac{w_n(w_n - w_0)}{w_0} \qquad (6.10)$$

EXAMPLE 6.1 An evaporator concentrates a feed of 20% solids to 80%. Calculate the effect of a 1% change in steam flow or feed rate on product concentration:

$$\frac{dw_n}{d(\sum V_i/F_0)/(\sum V_i/F_0)} = \frac{0.8(0.8 - 0.2)}{0.2} = 2.4$$

A change of 1% in either steam or feed rate can cause product concentration to deviate $\pm 2.4\%$ from the 80% specification.

The sensitivity is primarily determined by the concentration ratio w_n/w_0, as seen in the example. Sensitivities in the 2–5 range are quite common, and explain why evaporators are difficult to control. Either steam flow or feed rate is ordinarily manipulated to control quality, with the other setting throughput. But regardless of the choice, it is imperative that the two variables be coordinated.

Equation (6.8) can also be differentiated with respect to feed concentration. While feed concentration is not used for control, it can vary substantially, particularly when feed is supplied batchwise. It is therefore a source of upset to the plant:

$$dw_n/dw_0 = 1/(1 - \sum V_i/F^o) = w_n/w_0 \qquad (6.11)$$

From Eq. (6.11), it appears that the sensitivity of product concentration to feed concentration is even greater than its sensitivity to the steam–feed ratio. A variation of 1% will cause a 4% upset in the product from the evaporator in Example 6.1. Consequently, feed composition cannot be ignored as a source of disturbance.

There is a difference, however, in the manner in which upsets develop. A change in feed rate can be introduced stepwise by an operator moving a valve or setting a flow controller. The impact of this disturbance depends on the capacity of the process side of the evaporator. The product from a multiple-effect calandria-style evaporator may start to change concentration very slowly, requiring perhaps half an hour to reach a new steady state. But an evaporator designed for minimum holdup as in Fig. 6.3 will respond to the full extent of the disturbance following a delay of only 2–3 min. The latter is therefore more sensitive to disturbances and more difficult to control.

If the feed composition could change stepwise, the dynamic response of product quality would be essentially the same as to a change in feed rate. However, feed is usually accumulated in a tank, where at least some mixing exists. Consequently even a step upset in a stream entering the feed tank will be moderated somewhat before it reaches the evaporator. Furthermore, variations in feed concentration may not range as widely as feed rate.

A change in steam flow must pass through all the heat-transfer surfaces in series before its full effect on product quality will be felt. A cocurrent low-holdup evaporator will tend to respond more rapidly to a change in feed rate. In a countercurrent evaporator, steam is introduced at the point where product is withdrawn, improving its dynamic influence on quality.

Changes in steam flow can be caused by an operator, or by variations in steam supply pressure. In Chapter 3, a method for compensating a steam flowmeter for variations in supply pressure was presented. While this will be effective against small disturbances, large upsets are also to be expected, where pressure is insufficient to deliver the required flow. Under these conditions, there is no recourse but to curtail feed rate accordingly— otherwise product specifications may be violated.

2.   Single-Loop Controls

The most common control configuration for evaporators consists of a series of single loops, as shown in Fig. 6.8. Production rate is essentially set by steam flow—an increase will reduce the level in all effects, forcing their level controllers to admit more feed.

Product quality responds both directly and indirectly to product flow. If product flow is decreased, while the same rate of steam is maintained, solids concentration in the last effect will begin to increase immediately. However a rising level in the last effect will also begin to decrease flow from the next-to-last effect. This action will be passed sequentially upstream by the level controllers. As these rates decrease, concentrations in the upstream effects

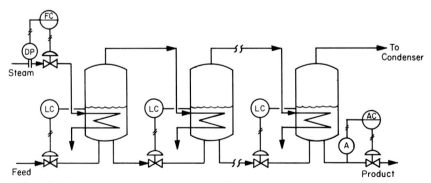

Fig. 6.8.   The most common control system for evaporators consists of a series of single loops.

begin to change. Eventually, the higher concentrations in the upstream effects reach the last effect, magnifying the original response.

The secondary responses arrive after a significant delay. The triple-effect evaporator described in Fig. 6.4 and Example 6.1 would have a product flow of 0.9 lb per pound of steam, in the steady state. If that rate were dropped to 0.89 lb, product concentration would rise from 80 to 80.4%, assuming no change in concentration from the second effect. However when the flow reduction is passed on to the second effect, it responds by concentrating its effluent, ultimately raising product concentration to 80.59%. When the flow reduction is passed on to the first effect, its concentration will rise, which eventually will bring product concentration to 80.67%.

The control system comprised of single loops is particularly sensitive to changes in steam flow. In order to change production rate, a change in steam flow must first upset product quality, and time must be allowed for the quality controller to then reach a new balance. During this interval, considerable off-grade product could be generated. Losses in steam pressure will have similar consequences.

3. Feedforward Controls

Evaporators designed for minimum liquid holdup cannot be controlled satisfactorily with uncoordinated single loops. Variations in steam and feed availability and feed concentration simply cause too great a loss in high-value product. For these evaporators, feedforward controls have been developed which coordinate changes in heat input, feed rate, and feed composition, to minimize deviations in product quality. They are based on the familiar material and energy balances.

For any given evaporator, the relationship between heat input and total vapor removal is essentially constant:

$$\sum V_i = kQ \tag{6.12}$$

where the proportionality k can be calculated according to Eq. (6.6) or is available as design information.

Then, using Eq. (6.3), the approximate heat-to-feed ratio for any given set of feed and product concentrations can be estimated:

$$Q/F_0 = (1/k)(1 - w_0/w_n) \tag{6.13}$$

However F_0 is a mass flow rate, which is not ordinarily measurable. Instead, either volumetric flow or orifice differential pressure are measured, both of which are sensitive to solution density. For the volumetric meter,

$$F_0 = F_{v0}D \tag{6.14}$$

where F_{v0} is the measured volumetric flow and D is the density of the feed. For an orifice meter,

$$F_0 = k_0\sqrt{hD} \tag{6.15}$$

where h is the measured differential pressure and k_0 is a meter factor.

The solids content of the feed is customarily measured as a function of feed density. The density measurement can then be used for two purposes at once. For the case of the volumetric flowmeter,

$$Q = (1/k)F_{v0}[D(1 - w_0/w_n)] \tag{6.16}$$

and for the orifice meter,

$$Q = (k_0/k)h[\sqrt{D}(1 - w_0/w_n)] \tag{6.17}$$

The bracketed terms in Eqs. (6.16) and (6.17) can be evaluated as functions of density:

$$f_v(D) = D(1 - w_0/w_n) \tag{6.18}$$

$$f_h(D) = \sqrt{D}(1 - w_0/w_n) \tag{6.19}$$

In order to evaluate these functions, the relationship between feed concentration and density must be known. Then a plot may be constructed of the

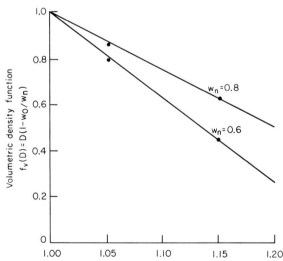

Fig. 6.9.   The volumetric density function for a typical corn syrup is remarkably linear with density.

appropriate density function against density. Figure 6.9 shows such a plot of $f_v(D)$ versus D for a typical corn syrup.

These functions come out to be surprisingly linear with density—even the orifice function, because the square-root curve does not change slope much in the region of 1.0–1.2. Consequently, they can both be represented by a linear model of the form

$$f(D) = 1 - m(D - 1) \tag{6.20}$$

where the slope of the line, which varies with product concentration, is adjusted by coefficient m. The intercept is always at (1, 1) for an aqueous system, because, as the feed density approaches that of water, all of feed must be evaporated.

If the heat input is to be manipulated for control, its set point is calculated by the feedforward system in response to variations in feed rate and composition as follows:

$$Q^* = (1/k)F_{vo}[1 - m(D - 1)] \tag{6.21}$$

A similar form exists for the orifice meter. In some plants, feed rate will be manipulated in response to variable steam flow and feed density:

$$F_{vo}^* = kQ/[1 - m(D - 1)] \tag{6.22}$$

Both of these systems are illustrated in Fig. 6.10. Note that the input from the feed density device is already $D - 1$ because the zero point on its scale would be the density of water; i.e., 1.0.

Figure 6.10 also includes dynamic compensators $f(t)$ on the flow inputs. They are provided to match the response of product composition resulting from a change in feed rate to that resulting from a change in heat input. A co-current flow evaporator of the low-holdup type responds faster to feed rate than to steam flow. Then if steam flow is manipulated to follow feed rate variations, a dominant lead is required. However, manipulating feed rate in response to steam flow requires a dominant lag. A lead-lag unit is typically provided with either system, adjusted as necessary to satisfy the process. A countercurrent evaporator may respond faster to steam flow, requiring a different combination of lead and lag settings.

4. Feedback from Product Quality

Feedforward calculations based on measurements of feed rate and density, and heat input, are not likely to be more accurate than $\pm 2\%$. Since an error of this order of magnitude can produce a deviation in product quality as great as $\pm 5\%$, feedforward by itself is insufficient for control—feedback must be added. This is not to admit that feedforward is not effective, however.

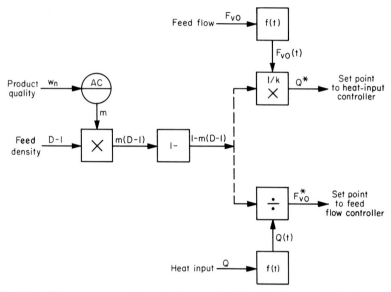

Fig. 6.10. The upper system is used when manipulating heat input, and the lower when manipulating feed flow.

Without it, the product-quality controller must provide all of the corrective action to rebalance the evaporator following a disturbance. But with feed-forward, only as much feedback is needed as necessary to correct for the *error* in the feedforward calculation. If the feedforward calculation is accurate to $\pm 2\%$, the feedback controller must only readjust its output to $\pm 2\%$ of the disturbance, rather than for the entire disturbance. And if feedforward action reduces the output change of the feedback controller by a factor of 50, its input, i.e., product quality deviation, will be reduced by the same factor of 50.

Feedback is properly applied to the variable in the feedforward system that most directly relates to product quality. In this case it is m, the slope of the density function. Any sustained deviation in product quality will cause the controller to readjust m, correcting the heat-to-feed ratio. Furthermore, an adjustment to the product-quality set point, which may be necessary when different grades are manufactured, will cause the controller to automatically recalibrate the feedforward model.

If a single grade of product is always manufactured, so that the slope of the density function is constant, feedback might be more properly applied to k. In this way, the system would be recalibrated for variations in heat losses and for flowmeter errors.

The feedback controller should not be tuned tightly, to avoid promoting oscillations. Because the feedforward system is capable of rapid correction for disturbances, the role of the feedback controller is reduced to the application of long-term correction.

A variety of measuring techniques is used to determine product quality. Density is the parameter usually detected, either by a nuclear gage or vibrating reed, with a hydrometer used for spot checking. Temperature correction may be required in some instances.

Because boiling-point elevation can be directly related to concentration, it has been used for quality control with mixed success through the years. There are problems associated with obtaining an accurate measurement. The bulb measuring the boiling point of the solution should be located at its surface to avoid errors associated with static or velocity head. It could also be in the vapor space, as the vapor is superheated to the boiling point of the liquid.

The dew-point of the vapor must be measured by cooling it sufficiently to cause condensation. This is done by locating the vapor temperature bulb in a recess which is exposed to a spray of water. Some adjustment may be required to achieve reproducible results.

The calibration of the instrument is also affected by absolute pressure, which changes both T and ΔH_v in Eq. (6.7). Compensation may be applied if pressure is uncontrolled. Rapid variations in pressure caused by cycling of the pressure controller or heat-input valve can promote different dynamic responses from the two temperature elements. Then ΔT may appear to cycle faster than composition can change. The feedback controller may reinforce this cycle, if it is tuned too tightly.

5. Trim Systems

When low-holdup evaporators were first introduced, grave difficulties were encountered in controlling the quality of their products. The conventional uncoordinated single loops were incapable of satisfactory regulation due to the dominance of dead time in the process. The absence of recirculation deprived these evaporators of the stabilizing capacity of older models.

In an effort to restore stability, the manufacturers added trim systems, at a cost both in capital and operating expense. A recirculation pump was added, with additional heat supplied by steam as shown in Fig. 6.11. Product quality responds more quickly to steam applied to the trim heater than to steam applied to the first effect. However, for economic reasons, the range of control over the trim heater must be severely restricted. Heat applied there is used only once where that applied to the first effect is used twice.

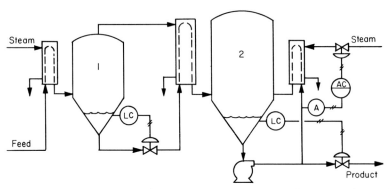

Fig. 6.11. A trim heater has been added to some evaporators to facilitate product-quality control.

Whenever variations in feed rate or density exceed a few percent, the trim steam valve can reach its limit of travel and control will be lost. As a result, the trim heater proves to be not only an expensive solution to the control problem but a less-than-satisfactory one as well.

With feedforward control applied to these evaporators, the role of the trim heater is largely eclipsed. In fact, control with feedforward is generally satisfactory without the trim heater, allowing it to be shut down and steam to be saved. Nonetheless, some products are able to benefit significantly from trim control. In these instances, the trim valve must be coordinated with the rest of the controls to be effective. A system incorporating the trim function is shown in Fig. 6.12.

The feedback control function is split into two components: a proportional-plus-integral controller is used to adjust the feedforward system, while the trim valve is manipulated with proportional-plus-derivative action. An important consideration is that the trim valve be directed to follow the

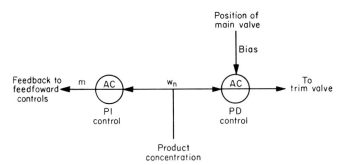

Fig. 6.12. The position of the trim valve is made to follow that of the main valve in the steady state.

feedforward inputs, as well as react to steam pressure changes. Both of these objectives can be accomplished by biasing the trim controller with the signal which operates the main valve as shown in Fig. 6.12. Any direction given to the main valve will then also move the trim valve by the same amount. Yet the trim valve is manipulated about that point to correct for derivations in product concentration.

While Fig. 6.11 only describes trimming action applied to a heater, it can also be applied to the feed. A small stream of feed can be injected into the last effect to achieve responsive control of product concentration. This practice is as inefficient as using a trim heater, in that this stream is only exposed to single-effect evaporation. With the advent of feedforward control, the trim stream is not required for most evaporators. Yet in those where some benefit may be realized, control over the trim valve is incorporated into the rest of the system just as described in Fig. 6.12. The bias signal is then the position of the main feed valve rather than the steam valve. The structure of the trim system is independent of whether steam or feed rate is manipulated by the feedforward system.

C. STEAM MANAGEMENT

The ability of evaporators to operate on low-pressure steam makes them very economical. In many plants, there is normally an excess of low-pressure steam so that a supply is assured for most operating conditions. While this situation is desirable from a production standpoint, it presupposes that low-pressure steam is wasted much of the time. As plant engineers act to reduce losses and find opportunities for use of low-pressure steam, the evaporators become central to the management problem.

1. Interruptible Systems

There is very little capacity in any steam system to store energy. If the flow of steam into a header is completely independent of header pressure, and the flow out is as well, header pressure shows no self-regulation, and can be expected to follow every change in load. The usual practice is to control header pressure by letting steam down from a higher pressure. This is not only inefficient because higher-pressure steam is more valuable but it cannot correct a condition of excess. Suppose that, on the average, the supply of low-pressure steam were equal to the demand. Sudden reductions in pressure could be averted by admitting higher-pressure steam. Yet this action would create an average surplus of low-pressure steam, requiring venting or condensing at other times.

A better solution to the problem is to allow the header pressure to fluctuate, thereby reducing both the admission of higher-pressure steam and venting.

These fluctuations can seriously upset steam users, however, including evaporators. Protection can be provided, in the way of steam flow controls. A change in pressure affecting steam flow would be immediately countered by the flow controller. But since the flowmeter calibration is also influenced by pressure, suitable compensation must be applied as described in Figs. 3.9 and 3.10.

In many cases, even pressure-compensated flow control is not enough. If there is a reduction in steam availability to a number of users having these controls, all valves will open, competing for the remaining steam. Unfortunately there will be times when all users cannot be satisfied. Those without compensated flow controls will take the brunt of the upset, but even some of those valves under flow control will be driven fully open. In an evaporator, loss of product-quality control will ensue, requiring diversion or blending of off-specification product.

There is a better solution to the problem. An evaporator that is not production-limited on the average can absorb fluctuations in steam supply, with its feed tank. Figure 6.13 shows an evaporator whose feed-tank level is controlled by setting steam flow. Actual heat input to the evaporator, as determined by the pressure-compensated flowmeter, is sent to the feed-forward system to be converted into the feed flow which will give the required product quality. That part of the control system is shown in the lower half of Fig. 6.10.

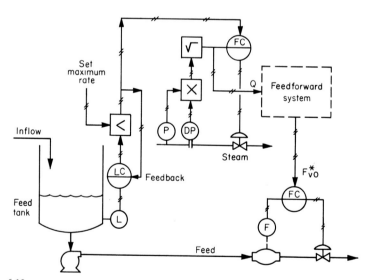

Fig. 6.13.   In this system, a reduction in steam supply will cause a loss in level control rather than quality control.

Should there be a reduction in steam availability to the point where flow control is lost, the system will automatically reduce feed rate. The reduction in feed will cause the level in the feed tank to rise. When steam again becomes available, the higher level will increase steam flow above what it was before the upset, and feed rate will again follow. There need only be enough steam available to meet the required production rate averaged over the time constant of the feed tank.

To maximize the capacity of the feed tank to absorb fluctuations in steam supply, the level controller should be set at a low value: e.g., 20%. In the steady state, tank level would then be held at 20%, with the remaining 80% free to accept inflow beyond what the evaporator is capable of concentrating. Should inflow increase sharply, or should insufficient steam be available, the proportional-plus-integral action of the level controller will increase the steam flow set point, possibly to the production limit. This limit should be adjusted to represent the maximum capability of the evaporator to deliver specification product. When steam supply is restored, the level controller may then demand maximum steam flow until the level returns to the 20% set point. This will return the system to its original capacity in preparation for another outage.

The proportional band of the level controller should be equal to its set point, e.g., 20%. Then a sharp reduction in level can be countered by an equivalent reduction in heat input and feed rate; zero heat input is then reached before the tank can be emptied.

Some newer evaporators will be equipped with steam-driven vapor compressors. Then steam turbines driving the compressors typically exhaust into the low-pressure steam header, which supplies conventional evaporators. An optimum work–heat balance will be attained when header pressure is controlled at its lowest practical value, by manipulating steam admission to the turbines. In Chapter 3, header pressure was controlled by admitting steam to turbogenerators, using the capacity of the electrical grid to absorb fluctuations in demand for low-pressure steam. With a vapor-compression evaporator, the capacity of the feed tank can be used to absorb variations in LP steam demand, as described above. However, to maximize the efficiency of the entire system, LP header pressure should be minimized, by control of the position of the most-open user valve, as shown in Fig. 3.8.

2. Steam Extraction and Admission

The evaporators described thus far have been independent of the rest of the plant except for their source of first-effect steam. In many newer installations, vapor from lower-pressure effects is extracted for other uses, such as space heating. Additionally, exhaust steam from low-pressure turbines or from flashing high-pressure condensate is being used to heat the latter effects.

This practice achieves economies and allows improved balancing of low-pressure steam supplies, but it also complicates the evaporator-control problem. Extraction or admission of vapor alters the system's energy balance and requires accounting. Figure 6.14 shows flowmeters installed on these lines to correct the heat-input signal to the first effect, in proportion to the rate of extraction or admission.

Consider a quadruple-effect evaporator where some vapor from the first effect is extracted for use elsewhere. From Table 6.1, the total vapor removed from the evaporator would be 3.52 times the steam *to* the first effect, less 2.71 times the vapor extracted *from* the first effect. Then the extracted flow would be multiplied by a factor of 2.71/3.52 or 0.77 before being subtracted from the main-steam flow signal. Similarly any steam combined with the vapors from the second effect would be multiplied by the factor 1.85/3.52 or 0.53 before being added to the main-steam flow signal. Correction for absolute pressure is probably not warranted in that these flows are ordinarily much lower than that of the main steam, and they have less influence over total vaporization.

Accounting becomes more difficult if an effect floats on a low-pressure header, which allows both extraction and admission through the same line. This requires a bidirectional flowmeter. A bidirectional orifice meter can be installed with reasonable expectations of accuracy, but the linearization of its output cannot be accomplished with a conventional square-root extractor. The differential-pressure transmitter would have a midscale output

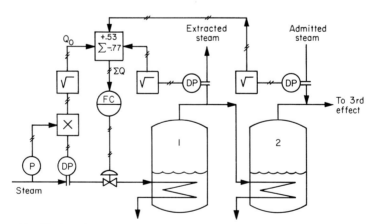

Fig. 6.14. The heat input to the first effect requires correction for steam extracted from or added to later effects.

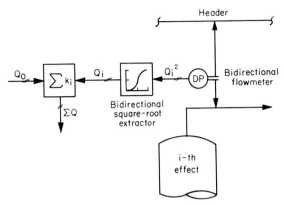

Fig. 6.15. A special characterizer is needed to linearize the signal from a bidirectional orifice meter.

at zero flow. Extracting the square root then requires the type of characterization described in Fig. 6.15. At zero flow, the output of the characterizer would also be midscale, requiring an offsetting midscale bias in the summing device.

3. Condenser Controls

With most evaporators, absolute pressure is controlled only at the last effect, whose vapor enters the condenser. If the vapor is free of contaminants and can therefore be used for boiler feedwater, a surface condenser may be used. If not, a direct-contact barometric condenser is usually applied. This allows a closer approach to cooling-water temperature in that the heat-transfer surface is eliminated, but the condensate and the cooling water are then mixed.

Regardless of the type of condenser used, a vacuum pump is also required to purge the vessels of air upon startup, and to remove infiltrated air during normal operation. Steam-jet ejectors are usually applied to this service, but mechanical vacuum pumps are also used. In most cases, the motive power for maintaining the vacuum is fixed, whether it is motor speed for a mechanical pump or steam flow to an ejector.

Absolute pressure is sensitive to both the rate of pumping and the rate of cooling, because both air and water vapor are drawn into the vacuum pump. The flow of water vapor depends on its partial pressure at the temperature of the condenser.

Consider a positive-displacement vacuum pump whose volumetric capacity is F_v. The molar flow rate of air M_a and water M_w which it removes varies with absolute temperature T and pressure p:

$$F_v = (RT/p)(M_a + M_w) \tag{6.23}$$

where R is the universal gas constant. The mole fraction water vapor at the inlet to the vacuum pump is its partial pressure divided by the total pressure. The partial pressure would be limited to the vapor pressure p° of the water in the condenser. Then

$$M_w/(M_a + M_w) = p^\circ/p \tag{6.24}$$

Combining (6.23) and (6.24) gives the absolute pressure in the condenser as a function of pumping capacity, air infiltration, and condenser temperature (which also determines vapor pressure):

$$p = p^\circ + RTM_a/F_v \tag{6.25}$$

Absolute pressure can be controlled by adjusting the rate of air infiltration into the vacuum pump, or by changing the condenser temperature with cooling water. If air is injected into the vacuum pump, more cooling water will be required to maintain a particular pressure. Then the lowest consumption of cooling water would be achieved when airflow is minimized. To conserve cooling water then, its flow should be manipulated to control pressure, without injection of air.

The next point to be considered is the optimum value of the controlled pressure. As the pressure is reduced, the boiling point of the product falls. The product then contains less sensible heat as it leaves the process, in proportion to the reduction in boiling point. From another point of view, less heat is needed to raise the feed to its boiling point as absolute pressure is reduced. The cost of energy representing the difference in sensible heat between feed and product is

$$\$_q/F_0 = c_q C_0(T_n - T_0) \tag{6.26}$$

where c_q is the unit cost of heat, C_0 is the heat capacity of the feed, and T_0 is its temperature.

To reduce the boiling point requires an increase in cooling-water flow, however. A heat balance on the condenser gives

$$Q_c = F_c C(T_n - T_c) \tag{6.27}$$

where F_c and C are the mass flow and heat capacity of the cooling water entering at temperature T_c.

For simplicity, let the heat flow be directly proportional to feed rate:

$$Q_c = KF_0 \tag{6.28}$$

The cost of cooling water is directly proportional to its flow:

$$\$_c = c_c F_c \tag{6.29}$$

When the two costs are combined, we have

$$\frac{\sum \$}{F_0} = \frac{\$_q}{F_0} + \frac{\$_c}{F_0} = c_q C_0 (T_n - T_0) + \frac{c_c K}{C(T_n - T_c)} \tag{6.30}$$

Differentiation of total cost with respect to T_n yields

$$(d\textstyle\sum \$/F_0)/dT_n = c_q C_0 - c_c K/C(T_n - T_c)^2 \tag{6.31}$$

When set to zero, the optimum value of T_n may be found:

$$(T_n - T_c)_{\text{opt}} = \sqrt{K(c_c/c_q)} \tag{6.32}$$

Here both C_0 and C were taken as 1.0 to simplify the results.

The essential message to be learned from (6.32) is that there exists an optimum temperature rise in the cooling water. Rather than maintaining a constant absolute pressure, condenser temperature rise should be controlled with cooling water flow as shown in Fig. 6.16. The optimum setting for the ΔT controller can be calculated from an energy balance on the evaporator, or adjusted to reach an observed minimum.

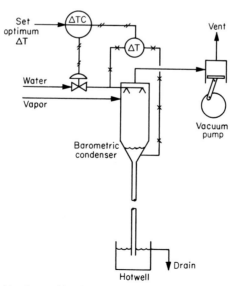

Fig. 6.16. The combined cost of heating and cooling will be minimized when cooling-water temperature rise is controlled at a constant value.

In most cases, the optimum absolute pressure seems to be higher than that maintained in a typical evaporator. Raising the pressure saves cooling water and reduces solids entrainment by decreasing velocities in all effects. While increasing product temperature does increase heat losses, it facilitates pumping of viscous products.

REFERENCES

1.   Spencer, G. L., and G. P. Meade, "Cane Sugar Handbook," 8th. ed., p. 699. New York, 1948.
2.   Stickney, W. W., and T. M. Fosberg, Treating chemical wastes by evaporation, *Chem. Eng. Progr.* (April 1976).
3.   Economical Evaporator Designs by Unitech, Bulletin UT-110, Ecodyne Unitech Division, Union, New Jersey, 1976.

Chapter

7

Drying of Solids

The operation of drying as described here consists in removing moisture from a solid material by exposure to unsaturated air or gas. This is to distinguish the process from evaporation where air is intentionally excluded. The single exception to this definition is the vacuum dryer, where moisture is removed by a vacuum system, with or without the application of heat. While this operation technically fits the definition of evaporation, it is included in this chapter because those familiar with vacuum drying will look for it here.

The rate of evaporation of moisture from the surface, and particularly the interior, of a solid particle is determined by a driving force, which can be described in a number of ways. If the process is viewed as one of mass transfer, the driving force can be related to the pressure difference between water on or in the particle, and the partial pressure of water in the surroundings. The process can also be treated as one of heat transfer, where the latent heat of vaporization is transferred to the moisture in the solid from the surroundings. The driving force for this heat transfer is then the temperature difference between the air and the solid.

Both of these considerations will be found useful in analyzing the relationships between the moisture contents, temperatures, and flow rates of solids and air. A psychrometric chart is useful in becoming acquainted with these relationships—therefore its role is presented first. Then the batch-drying process is examined, not for its economic importance but as an aid in understanding what happens in a continuous dryer.

A. ADIABATIC DRYING

A typical batch dryer is charged with wet material, through which a stream of heated air is passed until the product is acceptably dry. The operation is adiabatic if heat is applied to the air prior to its contact with the solid, rather than being applied to the solid itself. This distinction is important in that the operations of heating and drying are segregated, and therefore subject to independent analysis and control. Nonadiabatic drying is discussed at the end of the chapter.

1. Psychrometry

Psychrometry is the science of measuring the properties of moist air. There are several related properties, many of which bear directly on the drying process. In order to control the process effectively, certain of these parameters must be controlled in relation to others. Because the choice is crucial, the significance and means of measurement of these properties must be thoroughly understood.

Figure 7.1 is a simplified psychrometric chart for air and water at atmospheric pressure. The coordinates are absolute humidity in pounds of water per pound of dry air, and dry-bulb temperature. Absolute humidity may also be reported in terms of parts-per-million by weight or by volume, or grains of moisture per cubic foot of dry air.

The contour bordering the left side of the chart is the saturation line. Concentrations of moisture to the left of this curve cannot exist in a single phase—moisture will condense as dew, fog, or frost to the point of saturation. The temperature at which condensation begins is known as the "dew point." Each value of absolute humidity has a corresponding dew point at any given pressure. A dew-point measurement may then be used as an indication of absolute humidity, but only if the pressure is constant. As pressure is reduced, absolute humidity increases for a given dew point. The dew point in itself is significant in those processes where condensation needs to be avoided, as in ducts, filters, etc.

The triple point of water occurs at roughly 0.01°C. Below this point, water vapor will condense directly to the solid phase. Therefore dew-point temperatures below 0°C or 32°F are really frost-point temperatures.

Relative humidity is the mole fraction of moisture in the air, divided by the mole fraction at saturation for the same temperature and pressure. Another

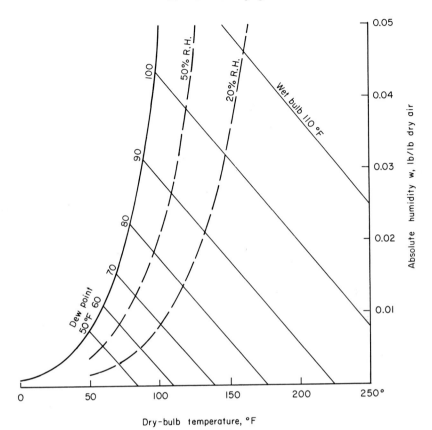

Fig. 7.1. The psychrometric chart is a road map for the drying process.

way to report relative humidity is the partial pressure of water in the air divided by its vapor pressure at a given temperature:

$$\% \text{R.H.} = 100(p_w/p_w^{\circ}) \qquad (7.1)$$

At saturation, the partial pressure and vapor pressures are equal, and the relative humidity is therefore 100%.

Relative humidity is most easily measured by the dimensional changes of such fibrous materials as hair, wood, or paper, in equilibrium with the air. It is the most important parameter to be controlled where this type of material is stored or used. For example, instrument charts and graph paper must be stored and printed in an atmosphere of controlled R. H., to maintain dimensional integrity. The moisture content of solid materials *in equilibrium* with air is related directly to its relative humidity.

Equilibrium prevails in long-term storage, but not in an actual drying operation. The atmosphere in a dryer departs substantially from equilibrium with the solids it surrounds, to create a driving force for drying. Consequently the factors affecting drying are distinctly different from those in an equilibrium situation, and different measurements are needed.

From a heat-transfer point of view, the rate of evaporation of moisture from a solid is directly proportional to the temperature difference between air and solid. The temperature of the air is its dry-bulb temperature. The temperature of the solid more closely approaches the *wet-bulb* temperature of the air, in that a wet bulb is in fact a wet solid from which moisture is being evaporated adiabatically.

Therefore wet-bulb temperature is a most significant factor influencing rate of drying. At ambient conditions, it is readily measured as the temperature of a surface from which an excess of moisture is evaporating into a large air space. It is most difficult to measure in hot air streams and within dryers, because conditions are created to promote rapid evaporation. The bulb surface is then easily contaminated with suspended, dissolved, and air-borne solids.

Nonetheless, it is possible to determine wet-bulb temperature, knowing dry-bulb temperature and absolute humidity, either from a psychrometric chart, or by trial-and-error calculation. The following procedure has been found to converge rapidly:

(1)   Estimate wet-bulb temperature T_{we}.

(2)   Calculate the absolute humidity at saturation (w_s) at the estimated wet-bulb temperature, using the known dry-bulb temperature T and humidity w_a:

$$w_s = w_a + (T - T_{we})/4100 \qquad (7.2)$$

(3)   Calculate the mole fraction of water vapor y_s corresponding to w_s:

$$y_s = 1/(1 + 0.62/w_s) \qquad (7.3)$$

(4)   Calculate the partial pressure p_s corresponding to mole fraction y_s by multiplying by total pressure p:

$$p_s = y_s p \qquad (7.4)$$

(5)   Using steam tables, read the temperature of saturation T_w corresponding to pressure p_s. If T_w is not in agreement with T_{we}, substitute T_w for T_{we} and repeat.

EXAMPLE 7.1   Calculate the wet-bulb temperature for air at 1200°F, having an absolute humidity of 0.025 lb water/lb dry air.

Estimate $T_{we} = 150°F$:

$$w_s = 0.025 + (1200 - 150)/4100 = 0.281$$
$$y_s = 1/(1 + 0.62/0.281) = 0.312$$
$$p_s = 0.312(14.7) = 4.59 \quad \text{psia}$$
$$T_w = 130°F$$

Second trial:

$$w_s = 0.025 + (1200 - 130)/4100 = 0.286$$
$$y_s = 1/(1 + 0.62/0.286) = 0.316$$
$$P_s = 0.316 \times 14.7 = 4.65 \quad \text{psia}$$
$$T_w = 131°F$$

Wet-bulb temperature varies much more with dry-bulb temperature and humidity at low values than at high values. This trend can be seen in the shape of the saturation curve. At saturation, wet-bulb temperature, dry-bulb temperature, and dew point coincide.

Lines of constant wet-bulb temperature are also lines of essentially constant enthalpy. This is consistent with wet-bulb temperature being described as adiabatic evaporation temperature. Air to be used in drying is first heated, which increases both its dry-bulb and wet-bulb temperatures at constant humidity, as shown in Fig. 7.1. Then as it contacts wet solids in an adiabatic dryer, its dry-bulb temperature falls, and its moisture content increases. If the temperature of the solids remains constant, the enthalpy of the air will also remain constant. It gives up sensible heat in return for latent heat of the water which joins it. Therefore the path which the air follows in proceeding through a dryer is along a line of constant enthalpy, which is also a line of constant wet-bulb temperature. Then the wet-bulb temperature of the air is uniform throughout the dryer. As a result, the temperature of the solid also tends to remain constant throughout the dryer.

In closing this discussion on psychrometry, it should be noted that the relationships are essentially those between water and an ideal gas. If the atmosphere were one of carbon dioxide, or hydrogen, or some other permanent gas, the same psychrometric chart could be used, amended for the molecular weight of the gas. Variations in atmospheric pressure can be treated in a similar manner. A reduction in pressure amounts to a reduction in the partial pressure of air, since that of water is determined by its vapor pressure at the given temperature.

2. Rate of Drying

Many factors affect the rate of drying. Air velocity, particle size, moisture content, and dry-bulb–wet-bulb temperature difference all contribute, along with the physical features of the dryer itself. At this point, the characteristics of the product need to be examined.

When a solid particle has absorbed all the moisture it can contain, its surface will appear wet. Upon placement in a drying environment, the surface moisture will begin to evaporate. If the dry- and wet-bulb temperatures of that environment are held constant, the rate of evaporation from the particle will remain constant, until dry areas appear on the surface. Because evaporation takes place entirely on the surface in this regime, its rate does not vary with the total moisture content of the particle. This regime is called constant-rate drying, as noted in Fig. 7.2.

With the first appearance of dry areas, the rate of drying begins to decrease. When the entire surface is dry, the rate is further retarded, as all moisture must diffuse from within. The rate continues to fall, eventually approaching zero, when the solid approaches equilibrium with its environment at a final moisture content w_e.

Figure 7.2 represents the rate-of-drying curve for an idealized particle. Most curves are rounded, and some display an inflection point in the falling-rate regime. Nonetheless, for most, there is a distinct and reproducible relationship between rate of drying and moisture content in the falling-rate regime.

The location of the critical point, at the transition between regimes, can vary with particle size. For example, the "Chemical Engineer's Handbook"

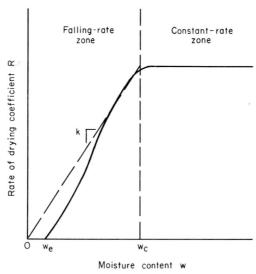

Fig. 7.2. At the critical moisture content w_c, dry areas begin to appear on the surface of a particle, and the rate of drying begins to fall.

(1) gives the critical moisture of sand as 5% for 50–150 mesh particles, 10% for 200–325 mesh, and 21% for smaller than 325 mesh. The pore structure of the sand is not likely to differ except for extremely small particles. However the amount of surface moisture which can be retained per pound of sand is higher for smaller particles, and the rate of drying is higher due to the larger surface area available. The increase in both surface area and critical moisture content leads to the conclusion that all particles approach the same falling-rate curve, regardless of size. The various curves would form an envelope as shown in Fig. 7.3. This relationship becomes important in establishing the performance of control systems encountering feedstocks having considerable variation in particle size.

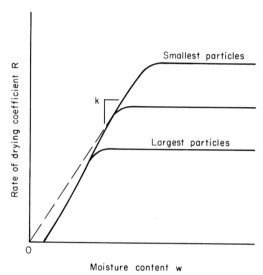

Fig. 7.3. Particle size seems to affect the critical moisture content without altering the slope of the falling-rate curve.

3. Batch Drying

The progress of a batch of solids exposed to air at constant conditions gives some insight into what is required for control. Consider the batch dryer shown in Fig. 7.4. A bed of wet solids is inserted into the dryer, and airflow is begun at a constant inlet temperature. A record of air temperature leaving is needed to inform the operator of progress, because he cannot sample the product without first opening the dryer.

The initial temperature of the solids may be below the wet-bulb temperature of the air. Then some sensible heat will be absorbed in raising the bed to the wet-bulb temperature. This will appear as a slight increase in air outlet

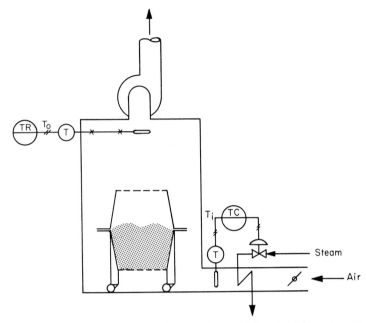

Fig. 7.4.   A bed of material is inserted into the dryer, and hot air is blown through it until the correct moisture content is reached.

temperature, after which a steady state is reached as shown in Fig. 7.5. During this observed steady state, the rate of drying is constant. A heat balance will bear this out. The temperature of the air is reduced proportionally to the evaporation of moisture:

$$GC(T_i - T_o) = F_w H_v \qquad (7.5)$$

where G is the mass flow of air, C the heat capacity of air, T_i the inlet dry-bulb temperature, T_o the outlet dry-bulb temperature, F_w the mass rate of evaporation, and H_v the latent heat of vaporization.

The actual value of T_o will lie somewhat above the wet-bulb temperature of the air. Because the solids temperature tends to approach T_w, T_o must exceed T_w to maintain a driving force for evaporation.

Let the rate of evaporation dF_w from a given particle be proportional to rate coefficient R, surface area dA, mass transfer coefficient a, and temperature elevation above the wet-bulb:

$$dF_w = aR \, dA \, (T - T_w) \qquad (7.6)$$

As the air passes the particle, its temperature falls:

$$GC \, dT = -dF_w H_v \qquad (7.7)$$

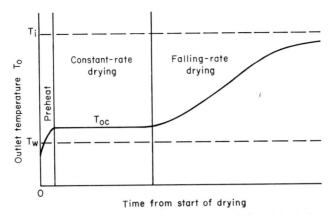

Fig. 7.5. Outlet-air temperature will vary between the wet-bulb and dry-bulb temperatures of the inlet air.

Combining these two expressions gives

$$dA = -\frac{GC}{aRH_v}\frac{dT}{T - T_w}$$

which, when integrated across the bed, yields

$$A = \frac{GC}{aRH_v}\ln\left(\frac{T_i - T_w}{T_o - T_w}\right) \qquad (7.8)$$

This allows the unknown factors of a, R, and A to be determined from an observation of temperatures.

Eventually the falling-rate regime is entered, signaled by an increasing outlet-air temperature. But as the rate decreases, as indicated by a falling $T_i - T_o$, the driving force increases, as indicated by a rising $T_o - T_w$. If the falling-rate curve can be approximated as a straight line passing through the origin, then R in Eq. (7.8) can be replaced with kw, having a maximum value of kw_c:

$$R = kw \leq kw_c \qquad (7.9)$$

Then the relationship between moisture content and temperature can be drawn by substituting (7.9) into (7.8):

$$w = \frac{GC}{akAH_v}\ln\left(\frac{T_i - T_w}{T_o - T_w}\right) \qquad (7.10)$$

Equation (7.10) can be evaluated during constant-rate drying, in which case w becomes w_c, and T_o becomes T_{oc}. If w_c and T_w are known, then the constants can be determined in lumped form:

$$\frac{akAH_v}{CG} w_c = Kw_c = \ln\left(\frac{T_i - T_w}{T_{oc} - T_w}\right) \tag{7.11}$$

Then Eq. (7.11) can be solved for the value of T_o at which the desired product moisture w^* is attained:

$$T_o^* = T_w + e^{-Kw^*}(T_i - T_w) \tag{7.12}$$

The procedure may be carried one step further by eliminating T_w in favor of w_c. Then T_o^* is found as

$$T_o^* = T_{oc}\left(\frac{1 - e^{-Kw^*}}{1 - e^{-Kw_c}}\right) + T_i\left(1 - \frac{1 - e^{-Kw^*}}{1 - e^{-Kw_c}}\right) \tag{7.13}$$

If w_c and K can be considered constants, Eq. (7.13) may be reduced to the form

$$T_o^* = K^* T_{oc} + (1 - K^*)T_i \tag{7.13a}$$

where K^* is a calibration adjustment sensitive to the desired product composition w^*, but insensitive to wet-bulb temperature.

4.   Controls for Batch Dryers

Equation (7.13a) can be implemented by a control system to terminate drying automatically when the product has reached its terminal moisture content. The system must memorize T_{oc} during a steady-state portion of constant-rate drying, for use in calculating T_o^* during falling-rate drying. Figure 7.6 illustrates how this is done.

The dryer is allowed to come to a steady state following startup. At the end of a prescribed time, the timer closes the solenoid valve, locking the

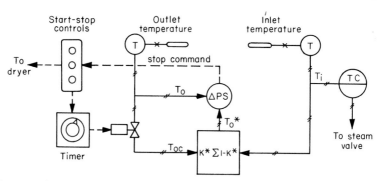

Fig. 7.6. The temperature corresponding to the desired product moisture is calculated in reference to the outlet temperature observed during constant-rate drying.

signal representing T_{oc} into the input of the pneumatic summing device. If the gain K^* of the summer has been properly set, the end-point temperature T_o^* will then be calculated from T_{oc} and T_i. When the measured temperature T_o reaches T_o^*, the differential pressure switch ΔPS will trip, terminating drying.

If K^* is set too low, T_o^* will be too high, approaching T_i, and the product will become too dry. If K^* is set too high, drying will be terminated while the product is still wet; drying must then be resumed, but without resetting the timer. Note that K^* will fall between zero where $T_o^* = T_i$, and one where $T_o^* = T_{oc}$.

By basing T_o^* on the observed value of T_{oc}, the system becomes completely independent of variations in air quality, as long as it does not change during the course of a batch. It is also relatively, but not completely, insensitive to the lumped constant K, as illustrated below.

EXAMPLE 7.2   Consider drying a batch of material having $w_c = 0.10$ to an end-point $w^* = 0.02$. Let inlet-air temperatures be 200°F dry-bulb and 93°F wet-bulb. Outlet-air temperature during constant-rate drying is 104°F. From Eq. (7.11),

$$K = \frac{1}{0.10} \ln\left(\frac{200 - 93}{104 - 93}\right) = 22.75$$

From Eq. (7.13),

$$K^* = \frac{1 - e^{-22.75(0.02)}}{1 - e^{-22.75(0.10)}} = 0.407$$

and, from Eq. (7.13a),

$$T_o^* = 0.407(104) + (1 - 0.407)(200) = 160.9°F$$

Let the next batch of material have only half as much surface area A. This will reduce K to half its original value. From Eq. (7.12), T_{oc} for the next batch is

$$T_{oc} = 93 + \exp[-22.75(0.10)/2](200 - 93) = 127.3°F$$

Based on this value of T_{oc} and the original K^*, T_o^* is

$$T_o^* = 0.407(127.3) + (1 - 0.407)(200) = 170.4°F$$

Using Eq. (7.10), the value of w reached at this temperature is

$$w = \frac{2}{22.75} \ln\left(\frac{200 - 93}{170.4 - 93}\right) = 0.028$$

If the original set point of 160.9°F had been retained for this batch, w would be

$$w = \frac{2}{22.75} \ln \left(\frac{200 - 93}{160.9 - 93} \right) = 0.040$$

Settings of K^* can be catalogued for the various products which must be dried in a given unit. Reproducibility between batches using these controls is found to be quite acceptable, virtually eliminating any need to reprocess material, or even to interrupt drying to sample the product.

B.   CONTINUOUS ADIABATIC DRYERS

Batch drying is restricted to low-volume processes such as in preparing pharmaceuticals, speciality foods, and in pilot-plant operations. But the insight gained in investigating batch drying facilitates the understanding of the continuous drying process.

There are two principal types of continuous adiabatic dryers—the fluid-bed, and the longitudinal dryer. The fluid-bed unit is featured by intense backmixing, such that the moisture content of the product is quite uniform throughout the dryer. In this respect, it is similar to the batch dryer whose product moisture is also uniform with respect to space, but varies with time.

The longitudinal dryer, on the other hand, having little or no backmixing, exhibits a gradient in moisture from inlet to outlet, in both the solid and gaseous phase. As a result, it is much more difficult to model than the batch and fluid-bed dryer. But being more common than they, its characteristics need to be evaluated as a basis for effective control.

1.   Fluid-Bed Dryers

A single-stage fluid-bed dryer appears in Fig. 7.7. A finely perforated bottom plate allows air to pass through, while retaining the suspended solids. Airflow is maintained at a rate that will fluidize the bed, but with a minimum of solids carryover into the exhaust. Differential pressure across the bed is an indication of its fluidization, and is therefore used to control air flow.

Wet feed enters on one side, and is thoroughly mixed with the contents. Dry product overflows through a port on the opposite side. The average moisture content of the bed is essentially that of the product leaving.

If the product leaves with more than the critical moisture, the dryer has no self-regulation. During constant-rate drying, air outlet temperature is as described in Eq. (7.11), related to the critical moisture content rather than the actual moisture content. Then w can vary with feed rate or feed moisture,

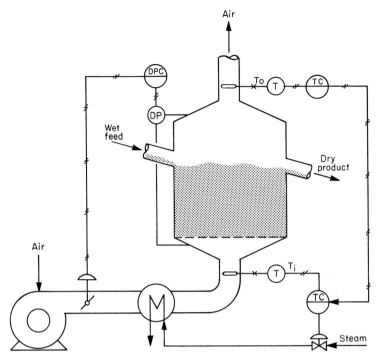

Fig. 7.7. A fluid-bed dryer is characterized by intense mixing, so that its contents are virtually uniform.

without having an effect on outlet-air temperature. In this case, the control system shown in Fig. 7.7 would not be able to regulate product moisture.

Normally, the product will leave with less than critical moisture, such that falling-rate drying dominates. Then, product moisture content w varies with air temperature T_o as described by Eq. (7.10). The control system shown in Fig. 7.7 will then provide partial regulation over product moisture. Increasing feed rate or moisture content will tend to raise w, which will increase the rate of vaporization, causing T_o to fall. Its temperature controller will react by increasing T_i enough to return T_o to its original value. The increased heat input will raise the rate of evaporation above what it was, but not enough to completely satisfy the increased load.

The deficiency in this control system can be explained by Eq. (7.10). At the higher evaporative load, T_i and T_w are both increased. If T_o remains constant, then w must increase to satisfy Eq. (7.10). In order to hold w constant at w^*, T_o must be controlled in relation to T_i and T_w, as described by Eq. (7.12). The difficulty in implementing Eq. (7.12), as a means to establish the set point for the outlet-air temperature controller, is the inability to measure wet-bulb temperature in the drying atmosphere.

Wet-bulb temperature can be estimated from dry-bulb temperature, and humidity or dew point. Figure 7.8 outlines the relationship over the range encountered in most fluid-bed dryers using indirect heat. (Another set of data, for direct-fired dryers, is given in Table 7.2.) Values of T_w may be taken from each curve in Fig. 7.8 and entered into Eq. (7.12) to produce a set of solutions for T_o^*, with e^{-Kw^*} as a parameter. A set of solutions is given in Fig. 7.9.

The curves themselves are quite complex and do not yield readily to mathematical modeling. Nonlinear characterization is possible for any given dew point, and this is generally sufficient, especially at the higher temperatures. There is less variation in T_w with temperature and humidity as temperature increases. Above 300°F, ambient humidity has almost no effect on the drying process. Actually the curves representing dew points of 0 and 80°F are extreme conditions, not often encountered in temperate zones.

There are two ways to control a fluid-bed dryer—either heat input or feed rate can be manipulated to regulate product moisture. If inlet tempera-

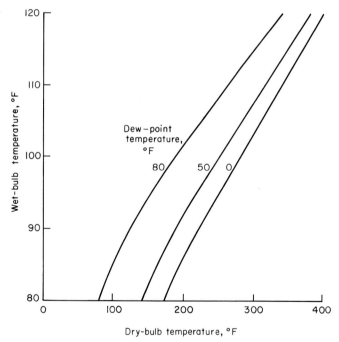

Fig. 7.8.   Wet-bulb temperature can be estimated from dry-bulb and dew-point temperatures.

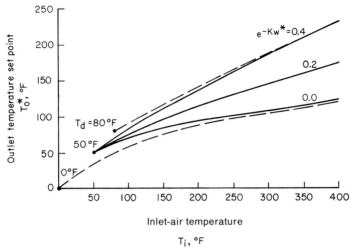

Fig. 7.9. Each of these curves is a contour of constant product moisture for a given inlet-air dew point.

ture is held constant, then, for a given ambient humidity, T_w will be constant and therefore T_o should also remain constant. Then a controller can hold T_o constant by manipulating feed rate. Variations in feed moisture which affect product moisture will affect T_o, and be countered by the controller acting on feed rate. If product moisture drifts due to changes in particle size, ambient humidity, etc., the operator need make only a minor adjustment in set point T_o^*.

Actually, feed rate is not often manipulated for moisture control. Because wet solids are difficult to convey, interruptions in flow are common. When a loss of feed occurs in a controlled situation, heat input must be reduced immediately to avoid overdrying the product. Furthermore, the only way production rate can be changed is by adjusting inlet-air temperature, which will require a corresponding change in T_o^*.

2. Inferential Moisture Controls

When heat input is manipulated, the outlet-air temperature must be adjusted as a function of T_i and T_w. Figure 7.10 shows a control system designed to approximate the curves given in Fig. 7.9. The curves are modeled as straight lines:

$$T_o^* = mT_i + b \tag{7.14}$$

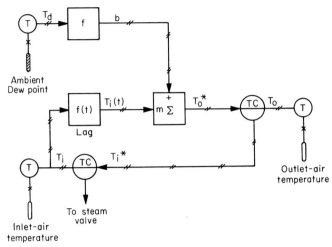

Fig. 7.10. Outlet-air temperature is set in a linear relationship with inlet-air temperature, biased by a function of ambient dew point, if necessary.

Bias b may be adjusted for variations in ambient dew point:

$$b = b_0 + f(T_d) \tag{7.15}$$

where b_0 represents the bias at zero dew point.

The bias correction $f(T_d)$ will vary with both T_i and e^{-Kw^*}. But since the correction is already small, adjusting it with these parameters seems superfluous. Figure 7.11 gives the functional relationship between b and T_d for an inlet-air temperature of 200°F and $e^{-Kw^*} = 0.2$. At higher temperatures, the correction will be smaller, and above 350°F it is not really needed at all.

To calibrate the system, e^{-Kw^*} must first be determined from plant design conditions, or observed values of T_o, T_i, and T_w:

$$e^{-Kw^*} = (T_o - T_w)/(T_i - T_w) \tag{7.16}$$

Then using tabular or calculated values of T_w at a nominal dew point (50°F is common), a curve similar to those in Fig. 7.9 may be plotted, following Eq. (7.12). Next, the best straight line is drawn through that curve, from which the intercept b and slope m may be determined. In operation, slope m should be adjusted to attain the required product moisture—increasing m will reduce moisture.

EXAMPLE 7.3   Consider a fluid-bed dryer having inlet air at 200°F dry-bulb and 55°F dew-point temperatures, and outlet air at a temperature of 130°F. Calculate m and b for the control system; anticipating that dew-point correction will be provided, calculate b_0.

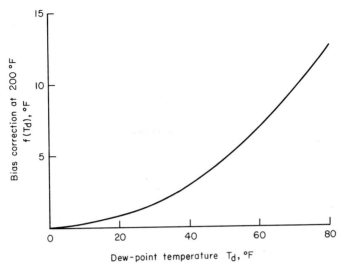

Fig. 7.11. This bias correction is only required for dryers having inlet-air temperatures below 350°F.

From Fig. 7.8, T_w is 92°F. Then from Eq. (7.16),

$$e^{-Kw^*} = (130 - 92)/(200 - 92) = 0.352$$

If the dryer is operated with $T_i = 150°F$, from Fig. 7.8, T_w is 82°F. Then T_o^* is, from Eq. (7.12),

$$T_o^* = 82 + 0.352(150 - 82) = 106°F$$

Next m is found between these two points:

$$m = \Delta T_o^*/\Delta T_i = (130 - 106)/(200 - 150) = 0.48$$

Then b may be found from the given point:

$$b = T_o^* - mT_i = 130 - 0.48(200) = 34°F$$

Figure 7.11 indicates that the bias correction applied to a 50-°F dew point is 4.8°F. Then

$$b_0 = b - f(T_d) = 34 - 4.8 = 29.2°F$$

Observe in Fig. 7.10 that increasing T_i causes an increase in T_o^*, whose controller will respond by further increasing T_i^*. This is a positive feedback loop, which is capable of destabilizing the process. If slope m is properly adjusted, measured outlet temperature T_o will respond *more* than T_o^* to a change in T_i. If m is too great, however, T_o^* could change farther than T_o can respond, causing the control system to diverge away from the steady state to either a high or low limit.

If T_o does not also respond *faster* than T_o^* to a change in T_i, cycling will tend to develop. To prevent this type of instability, a first-order lag $f(t)$ is introduced in the T_i signal path, as shown in Fig. 7.10. If its time constant is set too long, T_o^* will change too slowly with heat load, and product moisture will drift following load changes. Its setting should be reduced to the point where the first indication of cycling is observed. The dynamic contribution of this positive feedback loop to stability is essentially the same as the integral control mode (2). Therefore, the time constant of the first-order lag can be safely set near the integral time of the outlet-air temperature controller.

If drying is conducted in two stages, the product from the first stage is likely to have more than the critical moisture. Then the first-stage dryer is operating in the constant-rate regime, allowing its outlet-air temperature to act as a reference for the second stage. The procedure is the same as used to control the batch dryer, but the controls operate continuously. Figure 7.12 shows the control system implementing Eq. (7.13a), which was derived for batch drying. Note that both stages use the same heat source, so that there is a single value of T_i. This control system is insensitive to ambient humidity, and, like the batch-dryer system, compensates partially for variations in K as well.

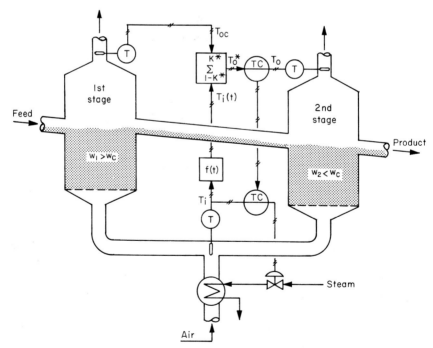

Fig. 7.12. The first stage of fluid-bed drying can act as a reference to control the second stage.

3. Longitudinal Dryers

There are many types of longitudinal dryers. The term is used here to identify those adiabatic dryers wherein product moisture varies from inlet to outlet. Then the rate of drying varies with position in the dryer, as a function of both air temperature and solids moisture.

It is possible to model these dryers mathematically, but obtaining a solution requires an iterative procedure. Differential elements of area must be integrated with respect to temperature, as in Eq. (7.8), and also with respect to moisture. The results of both must then equate to satisfy mass and energy balances.

An additional complication is that constant-rate and falling-rate zones may both be encountered, and their transition point tends to move with feed moisture. Most longitudinal dryers are fed cocurrently, allowing high inlet-air temperatures to be used, even with heat-sensitive products, because their temperature is limited to that of the wet-bulb. Heavy chemicals, cement, carbon black, and other products which are thermally stable and require a very low moisture content, are dried countercurrently.

There is a wide variety of dryer designs. Rotary dryers gradually convey the product from one end to the other by a slow rotating motion. Air-lift dryers convey fine or fibrous solids from the inlet to a cyclone separator at the outlet, where they are separated from the air stream. Pneumatic conveyence is also provided in some types of rotary dryers. In a rotating tray dryer, solids spill stagewise from top to bottom against an upward stream of moving air.

While not usually classified as longitudinal, the spray dryer is nonetheless characterized by similar properties. The fine slurry of feed material is sprayed counter to a moving stream of air so that it reaches the bottom in a dried condition.

Tunnel dryers are more complicated and present more control problems. They are typically divided into a number of zones through which the product is conveyed on a moving belt. Each zone may have independent control over inlet-air temperature, flow, and humidity. The first zones may be constant-rate zones. Belt speed, belt loading, and feed moisture are all variable. Inlet temperature profiles are often programmed to develop certain product surface properties, or to minimize hardening due to drying too rapidly. With all of these variations, it is difficult to recommend a general control approach for these dryers.

While the characteristics of longitudinal dryers differ substantially from those of the fluid-bed design, the same inferential moisture controls can be used for most. The principal distinction is that longitudinal dryers show a greater sensitivity of product moisture to temperature and to feed-moisture variations. A simplified model which seems to fit well replaces the exponential factor in Eq. (7.12) with a function of feed moisture w_f, if $w_f < w_c$:

$$T_o^* = T_w + (K_f/w_f)(T_i - T_w) \tag{7.17}$$

For a given dryer, K_f can be determined by inserting the design conditions for the other variables into Eq. (7.17).

Feed moisture is not usually measurable, but it can be estimated from a heat balance. The rate of heat transfer from sensible to latent heat is

$$Q = FH_v(w_f - w^*) = GC(T_i - T_o) \tag{7.18}$$

where F is the mass flow of dry solids into the dryer. Solving for w_f yields

$$w_f = w^* + (GC/FH_v)(T_i - T_o) \tag{7.19}$$

When (7.19) is substituted into (7.17), a quadratic in T_o results:

$$T_o = T_w + K_f(T_i - T_w)/[w^* + (GC/FH)(T_i - T_o)] \tag{7.20}$$

This relationship may be solved for selected values of T_i, T_w, and F to yield the required outlet-air temperature program to hold product moisture at w^*. Table 7.1 lists values of T_o corresponding to several sets of operating conditions for a typical dryer.

An examination of the figures in Table 7.1 indicates that $\Delta T_o^*/\Delta T_i$ for constant feed rate differs from that for constant feed moisture. Between the inlet-air temperatures of 356 and 302°F at 100% flow,

$$\left.\frac{\Delta T_o^*}{\Delta T_i}\right|_F = \frac{149.0 - 140.5}{356 - 302} = 0.157$$

Table 7.1

Outlet Temperature Required to Maintain Product Moisture at 13%

T_i (°F)	T_w (°F)	w_f (%)	F (%)	T_o^* (°F)
356	114	36.0	100	149.0
		42.5	80	143.6
		53.4	60	137.7
302	108	30.9	100	140.5
		36.1	80	135.9
		44.8	60	130.5
248	100	30.0	80	125.8
		36.5	60	121.1
		49.8	40	115.5

Yet between the same temperatures at 36% feed moisture,

$$\frac{\Delta T_o^*}{\Delta T_i}\bigg|_{wf} = \frac{149.0 - 135.9}{356 - 302} = 0.243$$

The higher slope is characteristic of fluid-bed dryers which are insensitive to feed moisture, while the lower is characteristic of longitudinal dryers operated at constant feed rate. This explains the field recalibration required for the spray dryer in Ref. (2).

The inferential control system for longitudinal dryers has inputs from temperature and feed rate:

$$T_o^* = b + m_T T_i + m_F F \tag{7.21}$$

For the dryer described in Table 7.1, m_T is 0.157 and m_F is 0.283°F/%.

4. Direct Firing

The high temperatures used in many dryers are achieved by the direct combustion of fuel, rather than by heat exchange with steam. Inlet-air temperature is controlled by manipulating the flow of fuel burned, with a constant flow of air.

The only complication this practice brings to the drying operation is the moisture added to the air as a result of combustion. Methane, the principal constituent of natural gas, forms 2 mol of water vapor for each mole burned. This amounts to about 0.94 lb of water per 10,000 Btu of heat. Assuming that the average heat capacity of the products of combustion is 0.24 Btu/lb-°F, each 1000 °F increase in temperature adds 0.0225 lb of water vapor per pound of air. Depending on the resulting air temperature, this could be more moisture than the air originally contained.

Fuel oil contains less hydrogen than natural gas does, resulting in a lower moisture increase—typically 0.01 lb per pound of dry air per 1000°F increase in temperature.

Knowing the moisture content of the combustion products for any inlet-air temperature allows an accurate prediction of the corresponding wet-bulb temperature, using the procedure given in Eqs. (7.2)–(7.4). Table 7.2 lists wet-bulb temperatures corresponding to given dry-bulb temperatures resulting from the combustion of methane with air having a dew point of 50°F. At the higher temperatures, ambient humidity has essentially no effect on wet-bulb temperature. Furthermore, wet-bulb temperature becomes less sensitive to moisture resulting from combustion. For example, at 1000°F the wet-bulb temperature resulting from the combustion of fuel oil is only 1.5°F below that resulting from the combustion of gas. At higher temperatures the difference is still smaller.

Table 7.2

Wet-Bulb Temperatures for the Combustion of Methane in Air Having a 50°F Dew Point

Dry-bulb temperature T_i (°F)	Wet-bulb temperature T_w (°F)	Dry-bulb temperature T_i (°F)	Wet-bulb temperature T_w (°F)
400	125	1200	159
600	134	1400	164
800	146	1600	168
1000	153	1800	171

Because wet-bulb temperature changes so little at high inlet temperatures, the contours of constant dryness shown in Fig. 7.9 are straight lines above 350°F. This makes the linear control system more accurate, as well as eliminating the complication of correcting for humidity variations.

One problem frequently encountered in direct-combustion systems is an inability to obtain a representative measurement of inlet-air temperature. Radiation effects can cause serious measurement errors, which change with sensor position and flame pattern. In these situations, a linear fuel-flow measurement has been substituted directly for inlet temperature in the inferential system of Fig. 7.10. This substitution assumes that airflow and ambient temperature are constant. Although they are not, the effects of their variation on the dryer model are not great. In any event, fuel flow is the best estimate of evaporative load, in the absence of an accurate inlet temperature measurement.

5. Feedforward Systems

Feedforward control is often added when there is a probability of rapid changes in feed rate. Loss of feed can create a fire hazard when air above 350°F is used to dry combustible materials such as wood fiber and grain.

Loss of feed is quite probable in dryers, due to the difficulty in conveying wet solids. In plants where other mechanical conveying equipment such as grinders, refiners, etc., connect directly at either end of the dryer, that probability is increased.

The feedforward input should be the mass flow of water into the dryer— this determines the difference between inlet- and outlet-air temperatures in the steady state. A steady-state heat balance on the dryer gives the following relationship:

$$GC(T_i - T_o) = FH_v(w_f - w) \qquad (7.18)$$

The set point for inlet-air temperature can then be estimated if feed rate and moisture are known:

$$T_i^* = T_o^* + (FH_v/GC)(w_f - w^*) \tag{7.22}$$

However T_o^* is already a function of T_i, according to Eq. (7.14). Combining this function with (7.22) gives

$$T_i^* = \frac{b}{1 - m} + \frac{FH_v}{GC}\left(\frac{w_f - w^*}{1 - m}\right) \tag{7.23}$$

Since feed moisture is not generally measured, and since all other terms except feed rate are constant, (7.20) can be simplified to

$$T_i^* = T_{i0}^* + K_F F \tag{7.24}$$

Factors T_{i0}^* and K_F can be estimated from a single set of stated operating conditions.

EXAMPLE 7.4   For the dryer described in Example 7.3, estimate values of T_{i0}^* and K_F, assuming that the conditions given were at 100% feed rate. From Eq. (7.23),

$$T_{i0}^* = b/(1 - m) = 34/(1 - 0.48) = 65°F$$

From Eq. (7.24),

$$K_F = (T_i^* - T_{i0}^*)/F = (200 - 65)/100\% = 1.35°F/\%$$

It is entirely likely that the intercept T_{i0}^* calculated above may be too low to allow prompt response of the dryer when feed is restored. But it cannot be elevated without a corresponding change in K_F, which would produce an incorrect reaction to a feed-rate change. Instead of elevating the intercept, a low limit should be placed on T_i^*, as required to allow rapid startup.

The inlet-air temperature controller should be protected against integral-mode saturation upon loss in feed, as described in Ref. (3).

6. Feedback from Product Quality

The inferential moisture-control systems described have a limited accuracy. They are based on assumptions of uniform particle size and drying curves, and a linear process model. Fluid-bed dryers, particularly of the two-stage variety, are likely to be controlled more consistently with these systems than longitudinal dryers, owing to their lower sensitivity to temperature and feed-moisture variations. Yet in every case, the inferential system will produce more consistent regulation over moisture than temperature control alone (2).

On-line calibration is required for most dryers, using the analysis of periodic samples of product as a guide. Some variation will be experienced in analysis of product samples, both due to analyzer inaccuracy, and sampling inconsistencies. These factors should be considered when weighing results.

Nonuniformity of particle size and moisture content are problems affecting both analysis and control. Large lumps of solids are difficult to dry, and only their outside surface may have a significant impact on temperatures. If these lumps are pulverized and subject to analysis, the average moisture content may have no bearing on conditions in the dryer. In situations like this, size reduction and classification should be provided to allow uniform drying and reproducible control.

If a reliable on-line analyzer is available, product moisture can be controlled with it. However the moisture controller should not manipulate either heat input or feed rate directly. The dead time in response of composition to these two variables is too long for the loop to be dynamically responsive. Instead, the moisture controller ought to adjust slope m of the inferential system described by Eq. (7.14). Temperatures respond much faster to upsets in feed rate, feed moisture, and air conditions than can product moisture. The inferential system can then correct for these upsets rapidly, if with limited accuracy. The moisture controller can then correct for long-term errors, based on product analyses.

C. MINIMIZING ENERGY USAGE

Most dryers are operated in such a way that their product is nearly always overdried. The purpose is to avoid violating specifications on the product, in the face of disturbances, and in the absence of on-line analysis. Therefore, the first goal of any dryer-control system should be to provide consistent regulation, which will allow a reduction of this operating margin.

Once having established quality control, attention can then be turned to energy-conserving techniques associated with airflow manipulation and heat recovery. Using sources of waste heat for drying is another important consideration, which requires effective control, to be acceptable to operating personnel. These subjects are considered in order of importance.

1. The Cost of Overdrying

There are several cost factors associated with overdrying:

(1) product giveaway
(2) fuel consumption
(3) production capacity
(4) air pollution
(5) fire hazard

The case for reducing product giveaway was covered under the subject of evaporators. While this is also an important consideration for dryers, fuel savings may be as significant in many cases. But both factors may be determined using the models already developed.

The air-pollution problem is not universal, but is characteristic of some products. Gypsum, for example, when overdried, is easily pulverized in the normal attrition of the dryer, and can send a cloud of dust into the exhaust system. A fire hazard exists for combustible products like wood fiber and grains. A sudden loss in feed can cause rapid overdrying of residual particles to the point of combustion.

Only a fracion of the heat input to a dryer is used to evaporate moisture. Its thermal efficiency may be stated as the ratio of heat absorbed in evaporation to the heat input. It can be calculated as the ratio of the reduction in air temperature during the drying process to the elevation in air temperature in the heating process:

$$\eta = (T_i - T_o)/(T_i - T_a) \qquad (7.25)$$

where T_a is ambient temperature. If the dryer in Example 7.3 uses air at an ambient temperature of 60°F, its thermal efficiency would be $(200 - 130)/(200 - 60)$, or 50%.

If the product is overdried, additional heat is supplied for evaporation, but more is used to raise the wet-bulb temperature and driving force. Each incremental reduction in product moisture requires a greater outlet-air temperature elevation to overcome the reduced drying-rate coefficient. Fluid-bed dryers will require substantially more heat to overdry than longitudinal units, because the entire contents of the dryer are lower in moisture. Longitudinal dryers, however, would have reduced moisture content at the product end, but not the feed end, so that their average rate coefficient would not change as much.

To estimate the additional heat required to overdry the product from a fluid-bed dryer, Eq. (7.12) can be used, along with a heat balance. Because of the exponential term in (7.12), and the dependence of T_w on T_i, a direct solution of heat input as a function of product moisture is unattainable. However, a set of numerical solutions may be obtained as illustrated below.

EXAMPLE 7.5   For the dryer described in Example 7.3, calculate the percentage increase in fuel consumption if the product is dried to 3% moisture instead of the specified 4%. The moisture content of the feed is 10%, and the ambient temperature is 60°F.

From Example 7.3,

$$e^{-Kw^*} = 0.352$$

Then

$$Kw^* = -\ln 0.352 = 1.044$$
$$K = 1.044/0.04 = 26.1$$

A heat balance using Eq. (7.18) gives

$$\frac{FH_v}{GC} = \frac{T_i - T_o}{w_f - w^*} = \frac{200 - 130}{0.10 - 0.04} = 1167$$

and the heat required to raise air at 60°F to T_i is

$$Q = GC(200 - 60) = 140GC$$

At $w = 0.03$,

$$e^{-Kw} = e^{-26.1(0.03)} = 0.457$$
$$T_o = T_w + 0.457(T_i - T_w)$$

From a heat balance,

$$T_i - T_o = (FH_v/GC)(w_f - w) = 1167(0.10 - 0.03) = 81.7$$

The last two expressions can be combined by eliminating T_o, giving

$$T_i = T_w + 81.7/0.543 = T_w + 150.5$$

The actual value of T_i must be found using Fig. 7.8 or a psychrometric chart:

$$T_i = 250°F$$

Then the heat required to raise 60°F air to T_i is

$$Q = GC(250 - 60) = 190GC$$

This represents an increase of 35.7% over the base case, although the increase in moisture removal is only 0.07/0.06 or 16.7% over the base case.

The relationship between heat requirements and product moisture is highly nonlinear. Figure 7.13 plots the relationship for the fluid-bed dryer in the above example. Substantial incentive exists to operate at as high a product moisture content as is acceptable to the user of that product.

Most dryers will be heat-input limited. Then, if heat input per unit feed can be reduced by controlling product moisture closer to specifications, more feed can be processed. In many cases, the increased productivity can be worth more than fuel savings or reduced product giveaway.

2. Optimum Airflow

Most dryers operate under conditions of virtually constant airflow. Actually, airflow is not often controlled at all, except for fluid-bed dryers where

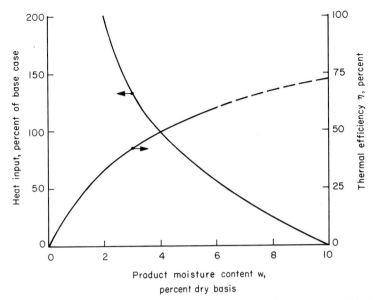

Fig. 7.13. The relationship between heat input and product moisture for the fluid-bed dryer is highly nonlinear.

it is important to maintain fluidization. One factor worth considering, then, is the effect of airflow on the economy of a dryer.

If airflow has no effect on mass or heat transfer, then increasing it will increase energy consumption substantially. This can be demonstrated by doubling G in Eq. (7.10) for the fluid-bed dryer described in Example 7.5. The improvement in K lowers T_i from 200 to 171°F, not enough to offset the increased flow through the air heater; the result is an increase in heat input of 59%.

However, in most systems increased airflow does improve the mass-transfer coefficient. If the relationship between a and G in Eq. (7.10) is linear, K then is unaffected by airflow. In this case, the net heat input does not change with G, higher flow rates simply reducing the required inlet temperature.

A real situation is somewhere between these extremes. Typically, in a turbulent flow regime, mass- and heat-transfer coefficients vary with the 0.8 power of flow. This would cause K to vary with $G^{-0.2}$. With this low exponent, airflow does not have a great effect on heat input, but the results are worth examining. Table 7.3 shows that reducing airflow from the base case of Example 7.5 results in improved thermal efficiency.

If this model is representative, then airflow should be reduced to some allowable *minimum*, which would then maximize thermal efficiency. However

Table 7.3

The Effect of Airflow on Heat and Work Requirements

Airflow (% of base)	Inlet-air temperature (°F)	Thermal efficiency (%)	Work input (% of base)
200	136	46	63
100	200	50	100
50	308	56	141

the higher temperatures needed may not be available. The work input to the dryer was determined by applying Eq. (1.10) to the heat input. This is seen to increase as airflow is reduced. Then the optimum work efficiency would appear to be reached at *maximum* airflow.

In actual practice, work efficiency may be important during the design phase, but not in operation, unless a choice of several heating media exists. If there is no choice, then apparently airflow is optimized when air-inlet temperature is as high as attainable with the given heating medium. For example, if steam at 320°F is available for the fluid-bed dryer in Table 7.3, airflow should be reduced as low as possible, while still retaining control of fluidization and product moisture. However, the dryer could operate more efficiently from a work standpoint, if a lower-temperature heating medium were provided.

An additional factor not considered in Table 7.3 is the increase in sensible heat of the product at lower airflows. As inlet-air temperature increases, wet-bulb temperature increases too. The temperature of the product must rise with it. This will result in a heat loss, and therefore a slight reduction in dryer efficiency as airflow is lowered.

Many dryers require a high limit on inlet-air temperature to avoid degradation of the product. If this temperature can be maintained at all times by manipulating the heating medium, then airflow can be manipulated to control moisture. The control system required for this configuration is simpler than those having a fixed airflow—Fig. 7.14 shows only two loops. If either valve reaches the fully open position, control of product moisture will be lost. Therefore their positions should be indicated and alarmed, as shown.

There are two limitations to this system. Outlet-air temperature is difficult to control by manipulating airflow, due to the variable dead time of the air within the dryer. A similar control loop on a heat exchanger could not be stabilized satisfactorily (4). The second problem is that outlet-air temperature should change somewhat with airflow.

Following the information presented in Table 7.3, calculations were made to determine the airflow and outlet temperature required to hold a constant

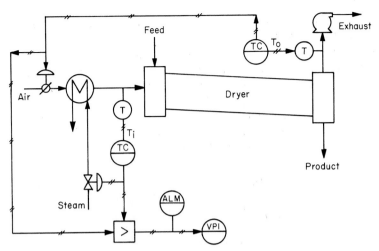

Fig. 7.14. Minimum airflow can be achieved if it is manipulated to control outlet temperature at a constant inlet temperature.

product moisture, given 308°F as the maximum inlet temperature. The results for the fluid-bed dryer used in the previous examples appear in Table 7.4. The change in T_o is small enough that it could possibly be neglected—it is certainly much smaller than the change required with constant airflow.

3. Recycling Exhaust Air

Some of the heat content in the exhaust may be recovered if the air is sufficiently hot and dry. This can be particularly attractive in batch drying, where outlet-air temperature rises and humidity falls as the end point is approached.

For heat recovery to be effective, without interfering with the primary objective of product moisture regulation, it must be properly controlled. Variations in humidity at the inlet of the dryer can substantially alter the

Table 7.4

Thermal Efficiency and Outlet Temperature for Airflow Manipulation at
$$T_i = 308°F$$

Evaporative load (% of base)	Airflow (% of base)	Thermal efficiency (%)	Outlet temperature (°F)
50	23	61	157
100	50	56	168
150	77	55	173

relationship between product moisture and air temperatures, particularly for low-temperature drying.

The variable which limits the amount of recycle allowed is humidity. It can be controlled successfully using a dew-point temperature measurement as shown in Fig. 7.15. For a batch dryer, little air would be recycled during the constant-rate period, when exhaust temperature is low and humidity is high. As drying proceeds, more heat recovery would be allowed. The combination of controlled dew-point and controlled inlet temperature is vital to the performance of the inferential moisture control system.

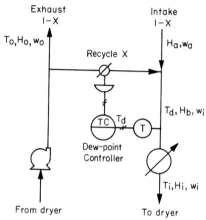

Fig. 7.15. The dew-point controller will allow the recovery of as much exhaust heat as practicable.

There would seem to be an optimum dew-point temperature for a given dryer. To evaluate the effect of dew point, the inlet-air driving force must first be determined, using the inlet–outlet differential corresponding to a given evaporative load:

$$T_i - T_w = (T_i - T_o)/(1 - e^{-Kw^*}) \qquad (7.26)$$

Then, using a psychrometric chart, or Eqs. (7.2)–(7.4), values of T_i and T_w must be found which satisfy their calculated difference, at the given dew point. Outlet-air conditions can then be determined from T_i and T_w.

Then a moisture balance is used to calculate the fraction X of exhaust air that is recycled:

$$X = (w_i - w_a)/(w_o - w_a) \qquad (7.27)$$

where w_i is the moisture content at the controlled dew point, w_o is the moisture content of the exhaust, and w_a that of the ambient air. Next, the enthalpy of the blend may be calculated from X:

$$H_b = XH_o + (1 - X)H_a \qquad (7.28)$$

where H_o and H_a are the enthalpy of exhaust and ambient air, respectively, as taken from a psychrometric chart. Finally, the heat input per unit of air is calculated as the enthalpy increase across the heater:

$$\Delta H_i = H_i - H_b \qquad (7.29)$$

where H_i is the enthalpy at T_i and T_w.

The results of these calculations for the fluid-bed dryer described in Example 7.3–7.5 appear in Table 7.5. Ambient air is assumed to have a 50°F dew point and a 60°F dry-bulb temperature. There does not appear to be a clearly defined optimum: ΔH_i drops sharply as dew point is first increased, but later improvements are marginal. In fact, much of the heat recovery at higher dew-point temperatures will be offset by an increase in sensible heat absorbed by the product at higher wet-bulb temperatures. This factor was not taken into account in preparing Table 7.5. The logical conclusion to be drawn from the table is that increasing dew-point temperature beyond 70°F is probably not worthwhile. However a similar analysis conducted on the same dryer at a 50% higher load indicated a significant improvement in heat recovery between 70 and 80°F dew-point temperatures. In this light, 80°F might be the preferred set point.

Table 7.5

The Effect of Recycling on Heat Economy

T_d (°F)	X	T_i (°F)	T_w (°F)	T_o (°F)	ΔH_i (Btu/lb)
50	0	200	92	130	34.7
60	0.161	201.5	94.5	131.5	31.9
70	0.343	205	98	135	28.7
80	0.457	210.5	103.5	140.5	28.8
90	0.573	216	109	146	27.2

The selection of optimum dew-point temperature is independent of ambient conditions. Considerably more heat recovery is possible in winter when ambient humidity is low, than in summer when it is high. The dew-point controller can regulate the amount of recycle to adjust for ambient changes. In cases of very high humidity, the recycle damper may be closed, preventing any heat recovery.

Actually, a dryer without dew-point control requires more energy during the winter than during the summer, because the ambient dry-bulb temperature is lower, as well as the dew point. Table 7.6 shows that, with dew-point

Table 7.6

The Effect of Dew-Point Control under Various Ambient Conditions

Ambient conditions		Heat input (Btu/lb)	
Temperature (°F)	Dew point (°F)	With ambient dew point	With 80-°F dew point
10	0	43.0	31.8
40	30	37.3	29.5
60	50	34.7	28.8
80	70	30.2	29.0

control, heat-input requirements are essentially uniform for all seasons. A 10-°F difference between ambient dry-bulb and dew-point temperatures was used to illustrate each case.

In addition to heat recovery, a controlled dew point offers improved regulation of product moisture when temperature control is used. Figure 7.9 shows that control of dew point at 80°F will eliminate the variability of the dryness contours with ambient conditions, and removes much of their curvature. Both factors will improve the performance of the inferential control system.

4. Using Flue Gas for Drying

In many installations, flue gas from incineration or other similar operations is available for use in drying. The difficulty in attempting to capture such sources of waste heat is the variability of the source. At times, there may be more heat available than needed, and possibly at temperatures above safe levels for drying. At other times, there may be insufficient waste heat or none at all.

The system shown in Fig. 7.16 is designed to use all available waste heat, tempered with outside air, as necessary to provide the required inlet-air temperature. Although airflow will tend to vary somewhat, the data given in Table 7.4 indicated that the inferential control model is not particularly sensitive to its changes.

At times, the flue gas may not be hot enough to reach T_i^*, the temperature needed for moisture control. Then the difference between T_i^* and T_i is used to set a fuel flow controller in direct proportion. The ratio of fuel flow to temperature deviation is adjusted by the operator to produce that same temperature elevation at a nominal airflow. It may be necessary to place a low limit on fuel flow to avoid flameout at the burner. This incremental elevation in inlet temperature above T_i^* should be corrected by a corresponding increase in the bias adjustment b.

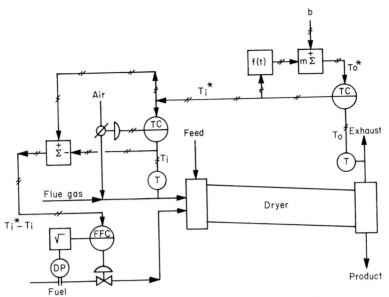

Fig. 7.16. Fuel is fired in proportion to the difference between actual inlet temperature and its set point.

D. NONADIABATIC DRYERS

There is a class of dryers that do not depend on a moving stream of heated air for evaporation. Drying is performed on contact with a heated surface. Water vapor is removed by natural convection or drawn off under a vacuum.

The most common dryer of this type is the steam-tube dryer (Fig. 7.17). It is a rotary dryer in which solids cascade against internal steam-heated tubes. Air flows counter to the feed from inlet louvers to an outlet stack, by natural convection. Because heat is applied throughout the entire length of the dryer, wet-bulb temperature increases from inlet to outlet.

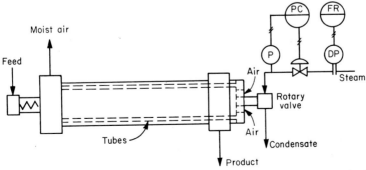

Fig. 7.17. Moisture is evaporated by direct contact with heated tubes in a steam-tube dryer.

Drum dryers are common in certain industries. Paper is dried in a continuous sheet in passing over a series of steam-heated drums. And flake food
products, chemicals, and soaps are dried from a paste which adheres to a
steam-heated drum. In all these examples, drying is accomplished by conductive heat transfer.

1. Drying by Conduction

When a heated surface such as a drum or tube is uncovered, the flow of
heat from its surface is governed by the low heat-transfer coefficient of the
surrounding air. If the same surface is then covered with a dry solid, such as a
sheet of paper or bed of dry fibers, the rate of heat transfer will not noticeably
change. However, should wet solids contact the heated surface, the rate of
heat transfer will increase substantially. In effect, the overall heat-transfer
coefficient is related to the average moisture content of the product in contact
with the surface:

$$U = f(\bar{w}) \tag{7.30}$$

If the heat-transfer coefficient U is controlled, then the average moisture $\bar{w}$
of the product in the dryer should also be controlled.

The heat flow Q from the steam-heated surface of area A is also related to
the temperature difference between the condensing steam T_s and the product
T_p:

$$Q = UA(T_s - T_p) \tag{7.31}$$

The load on the dryer is expected to change as a function of feed rate and
moisture content. The heat flow must also change in proportion. To increase
the heat flow requires an increase in U or T_s, or a decrease in T_p, if the
surface area does not change. Product temperature is not likely to decrease
as more heat is introduced, but in fact should increase as the wet-bulb
temperature increases.

Thus an increase in heat transfer must be caused by an increase in U or
T_s, or some combination of the two. If the increased load on the dryer is
brought on by a feed-rate change rather than a feed-moisture change, then
the moisture profile ought to remain the same. The only way that this can
be achieved would be to hold U constant, by increasing T_s in proportion to
the load.

In actual practice, steam temperature is not controlled, but is determined
by steam pressure in the dryer, which *is* controlled. In the absence of operator
attention, the steam pressure, and hence T_s, will remain constant regardless
of load. Then the only way that heat transfer can increase to accommodate
a rising feed rate is to raise U by allowing the average moisture content of

the product to increase. Conversely, a decrease in load will bring about a comparable reduction in product moisture.

An alert and experienced operator will adjust steam pressure when he perceives a load change. The indication he uses is the steam flow to the dryer. An increasing load will tend to increase U, causing steam pressure in the dryer to begin to fall. The pressure controller responds immediately by passing more steam, so that pressure is scarcely observed to change, while steam flow increases noticeably. The operator should react by raising the steam-pressure set point proportionately, thereby increasing its condensing temperature T_s. The set-point change will cause a secondary increase in steam flow: this is the result of operator intervention, and must not be construed as a another increase in load. This procedure can be mechanized to produce more accurate and responsive control of moisture than the operator is able to achieve.

2. Inferential Moisture Controls

The overall heat-transfer coefficient can be controlled inferentially by relating it to the observed steam flow rate:

$$Q = UA(T_s - T_p) = F_s(H_s - H_c) \tag{7.32}$$

where F_s is the mass flow of steam, H_s is its enthalpy, and H_c is the enthalpy of the condensate removed. Both T_s and H_c will change with steam pressure. Mass flow F_s is not directly measurable, but may be related to orifice differential h:

$$F_s = k_s\sqrt{h} \tag{7.33}$$

Combining (7.32) and (7.33) gives

$$k_s\sqrt{h}/UA = (T_s - T_p)/(H_s - H_c) = f(p_a) \tag{7.34}$$

The object is to relate h and absolute steam pressure p_a in a manner that will uniquely determine UA. To do this, values of T_s and H_c are obtained for selected steam pressures, and inserted into Eq. (7.34). A value of 180°F was used for T_p to generate the set of solutions appearing in Table 7.7. This value was selected because it was observed in operating steam-tube dryers— it appeared to change only ± 10°F with load. The steam supply is assumed to be saturated at 150 psig.

The last two columns from Table 7.7 are plotted against steam pressure in Fig. 7.18. Observe that orifice differential h is linear with pressure for a given value of U. In other words, steam pressure should be set in direct ratio to observed orifice differential, to hold U constant: The intercept is the saturation pressure corresponding to T_p, here slightly below atmospheric

Table 7.7

Solutions of Eq. (7.34) for Selected Steam Pressures

p_a (psia)	T_s (°F)	$H_s - H_c$ (Btu/lb)	$\dfrac{k_s\sqrt{h}}{UA}$	$\left(\dfrac{k_s}{UA}\right)^2 h$
30	250.3	977.1	0.072	0.0052
60	292.7	933.8	0.121	0.0146
90	320.3	905.3	0.155	0.0240
120	341.3	883.4	0.183	0.0333
150	358.4	865.4	0.206	0.0425

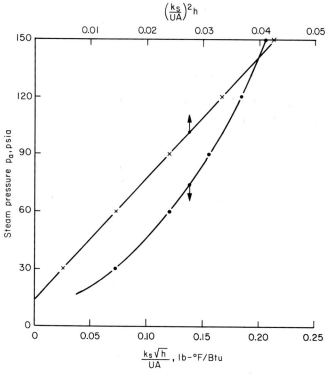

Fig. 7.18. Steam pressure is linear with orifice difice differential pressure for a constant heat-transfer coefficient.

pressure. In fact a gage-pressure measurement should be used, since it will accept the same changes in barometric pressure that affect T_p.

Then the line may be represented by the equation

$$p_g^* = (k_s/UA)^2 h \tag{7.35}$$

where p_g^* is the gage-pressure set point. The ratio of h to p_g^* is to be adjusted to provide the desired product moisture. An increase in the ratio setting will reduce U and therefore reduce the moisture content of the product. A control system embodying this concept appears in Fig. 7.19.

It is imperative that the meter coefficient k_s remain constant—otherwise U will shift in response. If the steam pressure at the point of metering is variable, compensation should be applied as shown in Fig. 7.19. The procedure is described in connection with Fig. 3.9.

Dynamic compensation in the form of a first-order lag is essential to the stability of the system. Only flow changes which reflect the true heat load should be converted into pressure set-point changes. Those flow changes resulting directly from set-point changes must be filtered, to attenuate positive feedback. The problem is essentially the same as encountered in the temperature loop of an adiabatic dryer. In essence, the dryer must be allowed to respond fully to an adjustment in steam pressure before another adjustment should be made.

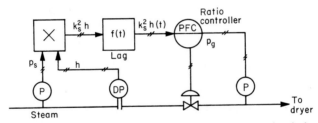

Fig. 7.19. Gage steam pressure is set in ratio to the lagged steamflow signal; the meter factor k_s should be compensated for supply pressure if it is variable.

Because the overall heat-transfer coefficient changes with average moisture content, the relationship shown in Table 7.7 and Fig. 7.18 is sensitive to feed moisture. As was the case with adiabatic longitudinal dryers, feed moisture cannot be conveniently measured, but can be inferred from a heat balance if a reliable feed-rate measurement can be made. First, let UA be a linear function of average moisture:

$$UA = k_f(w_f + w^*) \tag{7.36}$$

A heat balance relates w_f and F:

$$Q = FH_v(w_f - w^*) = F_s(H_s - H_c) \tag{7.37}$$

When combined with (7.34), a quadratic in h is formed:

$$\sqrt{h} = \frac{k_f(T_s - T_p)/k_s(H_s - H_c)}{2w^* + [k_s\sqrt{h}(H_s - H_c)/FH_v]} \tag{7.38}$$

Table 7.8

Steam Pressure Required for Constant
Product Moisture with Variable Feed
Moisture

p_a (psia)	w_f (%)	F (%)	h (%)
150	36.0	100	100
	39.9	80	84.9
	45.6	60	68.3
120	30.7	105	68.3
90	32.2	100	67.1
	36.0	78	56.3

From an initial set of conditions, k_f and k_s can be evaluated. Then h can be determined for various values of F and p_a. Table 7.8 gives selected combinations of conditions producing a product moisture of 10%, based on the initial set given in the first row.

From the first and last rows, the change in pressure with orifice differential for constant feed moisture can be determined:

$$\left.\frac{\Delta p}{\Delta h}\right|_{w_f} = \frac{150 - 90}{100 - 56.3} = 1.37 \quad \text{psi}/\%$$

This number is consistent with the data in Table 7.7. From the first and next-to-last row, the change for constant feed rate appears to be larger:

$$\left.\frac{\Delta p}{\Delta h}\right|_F = \frac{150 - 90}{100 - 67.1} = 1.82 \quad \text{psi}/\%$$

Finally the pressure correction with feed rate can be calculated from the two center rows:

$$\left.\frac{\Delta p}{\Delta F}\right|_h = \frac{150 - 120}{60 - 105} = -0.66 \quad \text{psi}/\%$$

The negative coefficient is consistent with the increased coefficient for constant feed rate: an increase in feed rate at constant feed moisture will bring about an increase in h. The inferential control system using a measurement of feed rate takes the form

$$p_g^* = \left.\frac{\Delta p}{\Delta h}\right|_F h + \left.\frac{\Delta p}{\Delta F}\right|_h F \qquad (7.39)$$

where the latter coefficient is negative.

3. Vacuum Drying

Certain heat-sensitive products are dried under a vacuum to reduce the temperature to which they are exposed. Some food products are of this category, along with high-quality soaps. The vacuum effectively removes all air from the system, so that the remaining atmosphere is essentially water vapor.

The driving force for evaporation may then be represented as the temperature difference between the product and the saturation temperature of water at that pressure. The rate of evaporation dF_w can also be expressed in terms of a differential-pressure driving force:

$$dF_w = a\,dA(p_w - p) \tag{7.40}$$

where a is a mass transfer coefficient, dA is an incremental surface area of solid, p_w is the partial pressure of water in equilibrium with the solid, and p is the absolute pressure in the dryer.

Soap is actually a solution of salts of fatty acids in water. Because of their high molecular weights (~ 300), the $10\text{--}12$ wt % of moisture in the final product exceeds 50 mol %. At these concentrations, Raoult's law applies:

$$p_w = p_w^\circ x \tag{7.41}$$

where p_w° is the vapor pressure of water at the temperature of the solid, and x is the mole fraction of water in the mixture. As x decreases, p_w decreases for a given temperature. Thus Eq. (7.41) is analogous to rate equation (7.9) in the falling-rate zone. If, instead of a solution, an emulsion or suspension is more representative of the mixture, the water will exert its full vapor pressure:

$$p_w = p_w^\circ \tag{7.42}$$

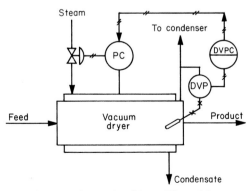

Fig. 7.20. The vacuum dryer can be regulated by a differential-vapor-pressure controller.

Then a drying-rate coefficient whose value varies with moisture content probably applies.

A study by the author on a soap dryer concluded that product moisture could be regulated by controlling $p_w^\circ - p$ as sensed by the differential-vapor-pressure transmitter described in Chapter 3. Its temperature bulb would be filled with water and inserted in the flow of product, with the pressure connection made to the dryer chamber. Steam pressure is the manipulated variable as shown in Fig. 7.20.

Dryer pressure is related to evaporative load, condenser water flow, and air leakage, as developed for evaporators in Chapter 6. Its control and optimization are as described in that context.

REFERENCES

1. Perry, R. H., and C. H. Chilton, "Chemical Engineers' Handbook," 5th. ed., p. 20-13. McGraw-Hill, New York, 1973.
2. Myron, T. J., Jr., F. G. Shinskey, and R. Baker, Inferential moisture control of a spray dryer, presented at the ISA Food Industries Symposium, Montreal, June 4-6, 1973.
3. Shinskey, F. G., Effective control for automatic startup and plant protection, *Can. Contr. Instrum.* (April, 1973).
4. Shinskey, F. G., Controlling unstable processes, Part II: A heat exchanger, *Instrum. Contr. Syst.* (Jan., 1975).

Chapter

8

Distillation

Distillation is a classic example of a process using energy to create order. Typically, the products of a distillation column are in the same physical state as its feed, and have the same net energy content. Therefore the energy used in the process simply passes through, increasing its entropy to reduce that of the products.

There are familiar methods available for measuring the efficiency of most processes such as boilers, compressors, engines, etc. These techniques are not directly applicable to distillation, in that order, rather than energy, is produced from energy. An entropy analysis appearing in this chapter gives a measure of efficiency and shows how much room there is for improvement.

Much energy can be saved by more effective quality control in reducing cycling and allowing operation closer to specification limits. But there are also energy-conservation techniques which apply to a column's heat-input and removal facilities. Minimizing the operating pressure of a column can improve its efficiency markedly. But control must be carefully applied to realize energy savings and to avoid such hazards as flooding and over-pressure.

Newer plants are making use of heat pumps and heat integration to reduce fuel consumption. These techniques are more demanding of controls because of the interactions they invite between competing units. They are illustrated by several applications to show how controls can be used both to save energy and facilitate operation.

A. OBJECTIVES AND PRINCIPLES

The foundation for this chapter has been established in an earlier work (1) by the author. There is not space in a single chapter to develop the subject of distillation control completely. Instead, the reader will be introduced to the objectives and principles of column operation sufficiently to present potentials for energy conservation. This will allow the benefits of improved quality control and other energy-conserving strategies to be converted quantitatively into anticipated savings.

1. Specifications on Product Quality

The entire purpose of distillation is to provide to a user a product of sufficiently high quality to perform satisfactorily in his service. Products that are sold must meet a set of guaranteed specifications listed in the buyer's contract or established by some authority such as the federal government.

Although only a few of the products from distillation columns are actually sold, most are given an arbitrary set of specifications, as if they were sold. In many cases, there is a logical explanation, in that the product must have a certain purity to avoid problems in downstream units of the plant. However, too often, specifications are arbitrarily applied to convenience inter-department transfer or accounting procedures.

When a product fails to meet specifications, some sort of penalty is imposed, depending on plant and market practices. The product could be sold as a lower-grade material, if there were a market for it. This penalty could be rather severe in that a product just failing to meet specifications may cost little less to manufacture than one which just meets them, and yet it must be sold for considerably less.

Another approach is to divert the off-grade material to a holding tank, where it can be blended with overpure product, to meet specifications. But blending is an irreversible process to which an energy penalty is attached. More energy is needed to make overpure product than can be saved by making that which is underpure. Blending the two to meet specifications will always consume more energy than making product of the required quality directly.

Below-grade product is often recycled to the feed tank. This practice is less efficient than blending with above-grade product—the entropy change increases with the difference between compositions, as indicated in Table 1.3. Furthermore, a column designed to process a certain feedstock cannot readily accommodate off-specification product as feed. Efficiency declines, and constraints of temperature, pressure, or flooding are easily approached. A general consequence of any of the above actions, except sale, is a reduction in plant capacity.

Because of the penalties associated with failure to meet specifications, columns are generally operated with a safe quality margin. Furthermore, composition records show that product purity tends to be less variable and therefore easier to control as it approaches 100%. This is borne out by sensitivity calculations. As a result, operators feel more comfortable when the product is overpure, and are not likely to force it closer to specifications, notwithstanding the economic incentives for doing so.

Specifications may be in terms of purity, impurities, or a combination of the two. Consider the depropanizer shown in Fig. 8.1. Its propane product may have to meet a 95% purity specification, and contain less than 2% isobutane. Or it could be stated as having to contain less than 2% isobutane and 3% ethane. Or it may have to meet the butane specification and have less than a certain vapor pressure, which would limit the content of ethane and lighter components.

The statement of specifications determines to a certain extent how the column is to be controlled. For example, the isobutane *impurity* in the propane product could be measured and controlled directly by manipulating product flow. To control *purity* is more complicated, because it cannot be measured with the same accuracy that it can be calculated. An analyzer accurate to $\pm 2\%$ of span could report propane purity only to $\pm 2\%$. But by measuring ethane and isobutane to $\pm 2\%$ of a 5% span, propane purity can be inferred to $\pm 0.2\%$. Therefore the sum of the impurities is generally used to control product purity.

Recognize that only the heavy key in a light product can be controlled at the column producing it. This would correspond to isobutane in the

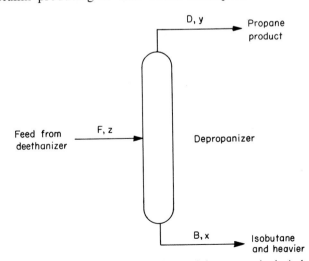

Fig. 8.1. The isobutane in the propane product and the propane in the isobutane product may both be controlled at the depropanizer.

propane product from the column in Fig. 8.1. If the lighter-than-light key in that product, i.e., ethane in the propane, is to be controlled, that must be done at a column farther upstream—in this case, a deethanizer. To control the ethane–propane ratio in the propane product, that ratio must be controlled in the bottom product from the deethanizer.

The light key in a heavy product may be controlled at the point of production. This would correspond to controlling the propane content of the bottom product from the depropanizer. Although this is not a final product, its propane content may be important in meeting the specifications of the isobutane final product downstream. The propane–isobutane ratio must be controlled to satisfy a propane specification on the isobutane product.

Some products may have no specifications at all. Their purity can then be optimized as a function of operating cost and product losses.

2. Material-Balance Relationships

As with any process, there is a binding relationship between quality and quantity, often overlooked in assigning control loops. An overall material balance may be written

$$F = D + B \tag{8.1}$$

where F is the mass or molar rate of feed, and D and B are those of distillate and bottoms products, respectively. One of these three streams can be used to control product quality, one establishes throughput, and the third closes the balance.

Similarly, a material balance may be written in terms of any component i:

$$Fz_i = Dy_i + Bx_i \tag{8.2}$$

where z_i, y_i, and x_i are the mass or mole fraction of component i in the associated stream. When (8.1) and (8.2) are combined, the product compositions appear as related flow ratios:

$$D/F = (z_i - x_i)/(y_i - x_i) \tag{8.3}$$

$$B/F = (y_i - z_i)/(y_i - x_i) \tag{8.4}$$

$$D/B = (z_i - x_i)/(y_i - z_i) \tag{8.5}$$

These expressions are not independent of one another. Because only one of the three flow rates can be used to control product quality, only one equation is useful.

Typically F and z_i are given as independent variables. Then either D or B may be selected to control the quality of its stream. Yet each of the above

Eqs. (8.3)–(8.5) has two unknowns: x_i and y_i. There then must be another relationship between x_i and y_i to define the system—that relationship is based on energy input.

Actually, the external material balance does not affect product compositions directly. As Fig. 8.2 shows, adjusting the rate of D or B does not impact directly on the column, but only affects the liquid level in the overhead accumulator or the column base. The internal liquid and vapor rates determine what the compositions will be.

However the internal rates are difficult to measure and control, and in some cases impossible. Vapor flow is not only determined by heat input, but also by feed rate and enthalpy. Liquid flow at any point is a function of the flow rate and enthalpy of both reflux and feed. Furthermore, in most columns, internal vapor and liquid rates are greater than product flows, such that their difference is more accurately controlled by the flow of a product.

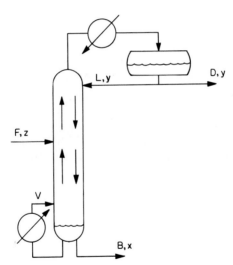

Fig. 8.2. Compositions within a column are related to internal vapor and liquid rates.

3. Separation as a Function of Energy

Separation s is defined as the ratio of light-to-heavy key concentrations in the top product, divided by the same ratio in the bottom product:

$$s \equiv (y_l/y_h)/(x_l/x_h) \tag{8.6}$$

By combining Eq. (8.6) with (8.3), written first in terms of light key l, and then heavy key h, a solution may be found for the four compositions given in

(8.6). The solution assumes no components lighter than the light key appear in the bottom product, and none heavier than the heavy key appear in the distillate. The result is quadratic in form:

$$y_h = (-b + \sqrt{b^2 - 4ac})/2a \tag{8.7}$$

where

$$a = (D/F)(s - 1)$$
$$b = sz_l - (s - 1)((D/F) - z_{ll}) + z_h \tag{8.8}$$
$$c = -z_h(1 - z_{ll}/(D/F))$$

Subscripts *ll* and hh represent all components lighter than the light key and heavier than the heavy key, respectively.

EXAMPLE 8.1   A depropanizer has the following compositions of feed and products, in mole fractions:

	Feed (z)	Distillate (y)	Bottoms (x)
Ethane and lighter (*ll*)	0.0097	0.030	—
Propane (*l*)	0.310	0.950	0.002
Isobutane (h)	0.088	0.020	0.120
n-Butane and heavier (hh)	0.593	—	0.878

Calculate D/F and separation. Estimate the effect of an increase of 0.01 in D/F on y_h:

$$\frac{D}{F} = \frac{0.31 - 0.002}{0.95 - 0.002} = 0.325, \qquad s = \frac{0.95/0.02}{0.002/0.12} = 2850$$

Let

$$D/F = 0.335$$
$$a = 0.335(2849) = 954$$
$$b = 2850(0.31) - 2849(0.335 - 0.0097) + 0.088 = -43.19$$
$$c = -0.88(1 - 0.0097/0.335) = -0.0855$$
$$y_h = \frac{43.19 + \sqrt{43.19^2 + 4(954)(0.0855)}}{2(954)} = 0.047$$

Reference (1, pp. 43–45) develops a correlation between the separation factor s and the average vapor-to-feed ratio V/F in the column:

$$V/F = \beta \ln s \tag{8.9}$$

where β is called the "column characterization factor." It is a function of the relative volatility α between the key components, the number of separation stages n in the column, and their overall efficiency E:

$$\beta = 3.50/[(nE)^{0.68}\alpha^{1.68}\ln\alpha] \tag{8.10}$$

There are two principal limitations to this correlation. The logarithmic model follows the actual points only in the range where

$$\ln s/(nE\ln\alpha) \le 0.522 \tag{8.11}$$

This is not a serious obstacle, because nearly all columns will be found to operate within this range.

The second deficiency is that the correlation contains no term relating to feed composition. A correction factor may be applied, however, using the composition effect on entropy appearing in Eq. (1.20). The correlation was derived for separation of equimolar binary mixtures, which have an entropy of mixing of $0.693R$ Btu/lb-mol-°F, where R is the universal gas constant, 1.987. Consider the feed to a column to be a binary mixture, with the light key and all lighter components lumped, and the heavy key and all heavier components lumped. Then V/F as calculated in Eq. (8.9) can be corrected by the ratio C_z of the entropy of the pseudobinary feed mixture to that of a 50–50 mixture:

$$C_z = \frac{(z_l + z_{ll})\ln(z_l + z_{ll}) + (z_h + z_{hh})\ln(z_h + z_{hh})}{-0.693} \tag{8.12}$$

EXAMPLE 8.2   Calculate the correction factor for the V/F correlation for the feed mixture given in Example 8.1.

$$C_z = \frac{0.3197\ln 0.3197 + 0.681\ln 0.681}{-0.693} = 0.904$$

Equation (8.9) is useful even when α and E are unknown or not uniform across the column. A set of conditions from an actual or accurately simulated column can be used to estimate β, from which new values of V/F can be determined as conditions are changed.

EXAMPLE 8.3   The column in Example 8.1 has a V/F of 1.60. Calculate the percentage reduction in V/F possible if the propane in the bottom product is allowed to increase to 0.3%.
From Eq. (8.9)

$$\beta = \frac{(V/F)_1}{\ln s_1} = \frac{1.60}{\ln 2850} = 0.201$$

The new separation factor is

$$s_2 = (95.0/2.0)/(0.3/12.0) = 1900$$

and the new V/F ratio is

$$(V/F)_2 = 0.201 \ln 1900 = 1.518$$

This amounts to a 5.1% reduction in V/F.

The energy Q required to perform the separation is directly proportional to vapor flow V:

$$Q = VH_v \tag{8.13}$$

where H_v is the latent heat of vaporization of the average mixture in the column.

One of the more valuable services of the correlation of energy with separation is the prediction of energy savings through an improvement in relative volatility. The relative volatility of most mixtures increases as temperature and pressure are reduced. If α is known at various conditions, Eq. (8.10) can be used to estimate the corresponding changes in V/F; a correction must be applied for changes in H_v with temperature in order to convert V/F into Q/F.

EXAMPLE 8.4   The relative volatility between propane and isobutane in the column of the previous examples is 2.01 at 120°F and 2.33 at 80°F. The heats of vaporization at these temperatures are 22,100 and 25,800 Btu/bbl, respectively. Estimate the percentage reduction in energy possible by operating at the lower temperature:

$$\frac{Q_{80}}{Q_{120}} = \frac{(\beta H_v)_{80}}{(\beta H_v)_{120}} = \frac{(2.01^{1.68} \ln 2.01)25,800}{(2.33^{1.68} \ln 2.33)22,100} = 0.752$$

A savings of 25% in energy usage is achievable.

4.   An Entropy Analysis

Energy conversion equipment such as boilers and compressors readily yield to efficiency analyses. Not so with mass-transfer processes, which use energy without producing any. Here, the entropy content of the energy is used to separate the feed mixture and thereby reduce its entropy.

For purposes of illustration, an entropy analysis follows on a column separating a 50–50 mixture of propane and propylene. These particular components are selected because their close boiling range makes separation difficult but also permits a wide selection of energy sources, including a heat pump.

Consider a column with 130 theoretical trays separating the feed mixture into two products, each 99% pure. Their relative volatility at 80°F is about

1.15, and their latent heat of vaporization averages about 3700 Btu/lb-mol:

$$s = \frac{0.99/0.01}{0.01/0.99} = 9801$$

$$\beta = \frac{3.50}{(130)^{0.68}(1.15)^{1.68}\ln 1.15} = 0.723$$

$$V/F = 0.723\ln 9801 = 6.65$$

$$Q/F = 3700(6.65) = 24{,}600 \quad \text{Btu/lb-mol}$$

The column is heated with steam at 212°F and cooled with water at 70°F, as shown in Fig. 8.3. Entropy of the feed and products were calculated using Eq. (1.20) and the gas constant R of 1.987 Btu/lb-mol-°F:

$$S_F = -1.987(0.5\ln 0.5 + 0.5\ln 0.5) = 1.377 \quad \text{Btu/lb-mol-°F}$$

$$S_D = S_B$$
$$= -1.987(0.99\ln 0.99 + 0.01\ln 0.01) = 0.111 \quad \text{Btu/lb-mol-°F}$$

The entropy change of the components per mole of feed is

$$\Delta S = S_D(D/F) + S_B(B/F) - S_F$$
$$\Delta S = 0.111(0.5) + 0.111(0.5) - 1.377 = 1.266 \quad \text{Btu/lb-mol-°F}$$

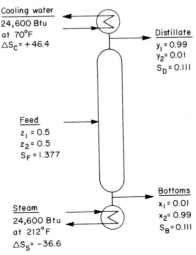

Fig. 8.3. The net gain in entropy of the surroundings is several times the reduction in entropy of the components being separated.

In the process of transferring 24,600 Btu/lb-mol of feed at 212°F, the entropy of the steam is reduced according to Eq. (1.32) by

$$\Delta S_s = \frac{Q}{T_s} = \frac{-24,600}{460 + 212} = -36.61 \quad \text{Btu/lb-mol-°F}$$

If the cooling water flows sufficiently fast that its temperature rise is very low, the same relationship may be used to estimate its gain in entropy:

$$\Delta S_c = \frac{Q}{T_c} = \frac{+24,600}{460 + 70} = +46.42 \quad \text{Btu/lb-mol-°F}$$

The heating and cooling system show a net gain of $46.42 - 36.61 = 8.81$ Btu/lb-mol-°F, compared to the net loss of 1.266 Btu/lb-mol-°F for the components being separated.

If a cooler fluid were used for reboiling, the operation would be more efficient in that the entropy increase would be smaller. This line of reasoning can only be carried to the point where the temperature difference between the hot and cold fluids approaches the difference in boiling points between the products. Their boiling point difference is about 16°F. The entropy increase for transferring 24,600 Btu from 86 to 70°F is

$$\Delta S = 24,600 \left(\frac{1}{460 + 70} - \frac{1}{460 + 86} \right) = +1.36 \quad \text{Btu/lb-°F}$$

Therefore, even when no temperature drop is taken across heat-transfer surfaces or in heat-transfer media, and no pressure drop is taken across the column, a net entropy increase is still sustained.

B. PRODUCT-QUALITY CONTROLS

The primary objective of any control system applied to a distillation unit is to meet product-quality specifications. Considering the penalties for failing to meet specifications, wide margins are typically set by the operators. As a control system is capable of reducing the variability in quality, operation closer to allowable limits can be achieved. Savings can be realized in the form of enhanced recovery of the specification product, or reduced energy consumption. Because these may prove to be the largest monetary incentives for improved controls in distillation, the relationship is worth examining in detail.

Distillation columns can be extremely difficult to control, for a variety of reasons. A typical column will have 4–6 control loops, which are largely interdependent. Loop selection is crucial to the performance of a column—the wrong combinations of manipulated and controlled variables can lead to extensive interaction and even instability. Most columns are easily upset by changes in the weather and disturbances in heating and cooling media as well. Proper loop selection can also help reduce that sensitivity.

An additional factor is the characteristically slow response of product analyzers to changes in manipulated variables. Here again, proper loop selection can help; still, most product-composition controls cannot respond quickly enough to overcome sudden changes in feed rate and composition. This is particularly true of close-coupled columns in series, where a disturbance can be propagated through the train, actually being amplified as it proceeds. In installations like this, feedforward control is mandatory if close control of product quality is to be achieved.

Within the scope of this section, it is impossible to cover the nuances of loop selection and interaction. Instead, emphasis is placed on the usual operating modes of columns, and the feedforward and feedback controls that best implement them.

1. Recovery, Energy, and Quality Control

There are two basic operating modes for distillation columns: maximum recovery and minimum energy. The maximum-recovery mode generally is applied to columns having one valuable specification product, and a less valuable byproduct. In this situation, operating cost will usually be minimized if the loss of valuable product into the byproduct stream is minimized—this constitutes maximum recovery of valuable product. This objective can only be achieved by using as much energy as allowed, because the separation factor between the products must be maximized. Boilup is then controlled by setting steam flow to the reboiler at some recognized upper limit. One way of determining the upper limit is to measure differential pressure across the column. As flooding is approached, the differential pressure usually rises sharply. Then the maximum heat input can be maintained by controlling column differential pressure just below the value where flooding is known to develop. This technique is particularly useful when the heating medium is oil or hot water, whose flow is not uniquely related to heat input.

If the two products are similar in value, improving recovery of the more valuable might result in less monetary savings than the additional cost of energy required. In this situation, economics would favor minimizing energy consumption. To achieve this goal, both products should be held as closely as possible to specifications, which will minimize their separation factor. This operating mode requires quality-control loops on both products.

In a few isolated cases, the cost of product lost through reduced recovery is comparable to the cost of the energy required to improve that recovery.

It may then be possible to arrive at a set of conditions that results in a minimum operating cost. Typically one product should be held at specifications, while the other is purer than required, by a stipulated amount—its composition is then optimized. This situation is covered in Refs. (1, pp. 325–345) and (2).

2. Feedforward Controls

A column can be protected against upsets in feed rate and feed composition by applying feedforward control (1, pp. 236–242). If both products are to be quality-controlled, the separation factor must be constant. One feedforward control loop in this system would then manipulate heat input Q in response to changes in feed rate:

$$Q^* = F(t) H_v \beta \ln s^* \tag{8.14}$$

where s^* is the separation corresponding to product-quality set points, and (t) indicates dynamic compensation of measured feed rate. If H_v and β are constant, (8.14) can be simplified to

$$Q^* = K^* F(t) \tag{8.15}$$

One of the material-balance equations such as (8.3) can be used to determine the second manipulated variable:

$$D^* = F(t)(z_I - x_I^*)/(y_I^* - x_I^*) \tag{8.16}$$

This equation would apply particularly well to the depropanizer of Example 8.1, where propane purity y_I had to be controlled, along with the propane content x_I in the bottom product. As before, the asterisk denotes set points rather than measured variables.

When x_I amounts to less than 1%, little accuracy is sacrificed by simplifying (8.16) to

$$D^* = m F(t)(z_I + z_{II}) \tag{8.17}$$

where m is an adjustable coefficient. For the column in Example 8.1, m is calculated to be 1.016, indicating there is little error encountered in using (8.17).

Some columns are not exposed to rapid or large changes in feed composition, in which case incorporating an analyzer to measure z_I and z_{II} cannot be justified. These terms can then be eliminated, with coefficient m enlarged to encompass them. Coefficients m and K^* are adjusted by feedback controllers to keep product qualities at their set points in the steady state. Figure 8.4 describes the control system.

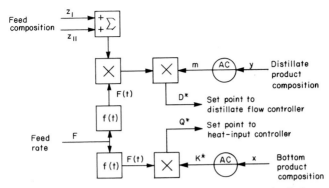

Fig. 8.4. This feedforward system applies to the minimum-energy mode of column operation.

Dynamic compensation $f(t)$ usually takes the form of dead time plus lead-lag action. If the feed contains no vapor, then the dynamic compensators are identical, with only dead time and lag required. In essence, the effect of a feed-rate change proceeds toward the column base. The counteracting change in heat input must be delayed until the effect of the feed change is felt at the column base. Very little time then elapses before the adjustment to heat input is felt at the top of the column. Therefore the distillate flow change can be made at essentially the same time as the heat-input change is made.

Any vapor in the feed tends to propagate directly to the top of the tower, however. Therefore if the feed is vaporized, even partially, the dynamic compensator manipulating distillate flow will have very little dead time, and some lead action will be required. A discussion on dynamic compensation and its adjustment is given in Ref. (3, pp. 211–219).

If the bottom-product flow is substantially smaller than the distillate, the accuracy of the material-balance calculation can be improved by manipulating B instead of D. Equation (8.4) or a simplification of it would be used. Feedback adjustment of m should come from the bottom-product composition controller. Base level control then must be transferred to heat input, and distillate would have to control the level in the overhead accumulator. Reflux flow would take the place of heat flow, to control distillate composition. Interaction between the two quality-control loops should also be considered in arriving at the most effective configuration, according to Ref. (1, pp. 306–311).

When heat input is fixed at some upper limit to maximize recovery, V/F will vary with feed rate, and so will separation. Then for a controlled y, x will vary with F. As a consequence, D is no longer linear with F, requiring a modification to the material-balance equation. Table 8.1 lists the flow rates

Table 8.1

**Distillate versus Feed Flow for
Constant Boilup and Distillate Composition**

F	V/F	x	D/F	D
1.0	5.45	0.05	0.479	0.479
0.7	7.79	0.0038	0.503	0.353
0.4	13.63	0.0002	0.505	0.202

and bottom composition for several feed rates to a propane–propylene column, having a fixed rate of boilup, and distillate purity controlled at 99%. The points of D versus F are plotted in Fig. 8.5.

A simple parabolic model can be fitted to the curve across the normal operating range:

$$D^* = mFz - kF^2 \tag{8.18}$$

where m is the output of the composition feedback controller and k is an adjustable coefficient which determines the nonlinearity of the curve.

Dynamic compensation of this system differs markedly from those which manipulate heat input. Lacking the rapid response afforded by changes in boilup, an additional feedforward loop is required to overcome the lag in the overhead accumulator. Otherwise, adjustments to D will only affect the liquid–vapor balance in the column *after* the accumulator level controller makes a corresponding change in reflux. With the forward loop given below,

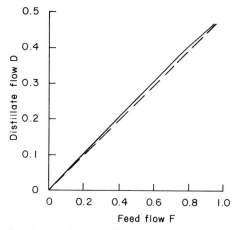

Fig. 8.5. When heat input is fixed, distillate flow varies nonlinearly with feed rate.

reflux is forced to change as soon as distillate changes:

$$\acute{L}^* = m_L - k_L D \qquad (8.19)$$

where m_L is the output of the accumulator level controller. If k_L is set at unity, L will respond directly to D, essentially eliminating the time lag of the accumulator. If k_L is too low, reflux will respond primarily to the output m_L from the level controller, with its associated lag. But if k_L exceeds unity, reflux will change more than distillate, giving the effect of lead-lag action. A control system embodying both Eq. (8.18) and (8.19) appears in Fig. 8.6.

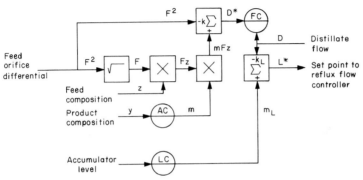

Fig. 8.6.   When heat input is fixed, a nonlinear model is needed, with dynamic compensation provided by the accumulator.

3.   Feedback from Product Quality

Product quality can be determined on the basis of a physical-property measurement, or by analysis. In the former category are measurements of boiling point, vapor pressure, density, and refractive index, with boiling point being the most common. These measurements are only indicative of true composition of binary systems, and even then are subject to errors caused by changes in temperature or pressure. In some cases, however, it is the physical property itself, rather than composition, which is to be controlled. This is particularly true of complex mixtures such as petroleum fractions, which are to be used as is, rather than be separated into their components.

Any measurement which can be inserted directly into the process, such as a temperature bulb, is much more responsive than one which requires a sample to be withdrawn. This is the principal failing in analyzers that are used for column control. A sample must be withdrawn, transported to the analyzer, conditioned, and analyzed, which may take several minutes. Therefore a typical period of oscillation of a feedback loop containing an analyzer would be 60–90 min, compared to 20–30 min for a temperature-control loop on a column. Reference (3, p. 102) indicates that the performance of a control loop, as measured by the time integral of deviation from set point, varies with the square of the dead time in the loop. Since the period is directly proportional to the dead time, an analyzer loop sustains perhaps nine times the integrated deviation that a temperature loop would on the same process.

But where temperature measurements lack the sensitivity and accuracy required by a particular separation, an analysis is required. An effort should be made to keep sample lines as short as possible, to transport vapor rather than liquid to the analyzer, and to minimize the response time of the analyzer. Sharing an analyzer between two columns may be economical, but it can be detrimental to control. Cascading an analyzer controller to a temperature controller can provide the accuracy of the analyzer with the responsiveness of temperature control.

The response of a loop containing a sample-and-hold device such as a chromatographic analyzer may be improved by using a sampled-data controller. Rather than operating continuously on progressively older information, the controller is placed in automatic only for a short interval after receiving a new analysis. It moves its output proportional to the deviation, and then is returned to manual, holding its present output to wait for another analysis.

An investigation of the steady-state gain of distillate composition to distillate flow reveals a variability:

$$dy_h/d(D/F) \propto y_h/x_l \qquad (8.20)$$

As y_h decreases, it responds less to manipulation in D, and, as it increases, it responds more. This characteristic reveals itself in a nonsinusoidal oscillation having flat valleys and sharp peaks.

This nonlinearity has been successfully compensated by dividing the composition measurement by itself plus the set point; the result is then applied as the controlled variable to a controller having a set point of $\frac{1}{2}$. The deviation developed at the controller is then

$$e = y_h/(y_h + y_h^*) - \tfrac{1}{2} \qquad (8.21)$$

The gain of e with respect to y_h now varies inversely with y_h, as required to provide a uniform loop gain. And when $e = 0$, $y_h = y_h^*$. This compensation appears in Fig. 8.7. The other composition loops are troubled by the same variability, and may be similarly compensated.

If the distillate is a final product with a purity specification, set point y_h^* should be calculated from the purity set point y_l^* and a measurement of the

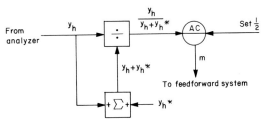

Fig. 8.7.   This system compensates the feedback loop for variations in process gain with composition.

uncontrollable impurity y_{ll}:

$$y_h^* = (1 - y_l^* - y_{ll}) \le y_h^{**} \tag{8.22}$$

The last term in (8.22) represents the impurity limit set on the controllable component. The system required to implement this function is shown in Fig. 8.8. In the event that optimization is possible, the optimum set point y_{ho}^* calculated for the controllable impurity would also be introduced into the selector as shown.

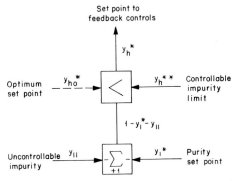

Fig. 8.8.   This system is devised to generate a set point which will satisfy the more stringent of two specifications, plus optimization, where applicable.

For an intermediate product, the ratio of the key components should be controlled. The set point for the key impurity would then be generated from a measurement of the other key, as

$$x_l^* = r^* x_h \tag{8.23}$$

where r^* is the desired ratio of the two components in the final product whose principal constituent is x_h.

C. ENERGY-CONSERVATION TECHNIQUES

The second half of this chapter is devoted to energy conservation and energy recovery. There is a subtle but important distinction between them. By energy conservation is meant that the separation itself is performed with a reduced consumption of energy, by being operated more efficiently. Energy recovery entails the reclamation of heat either from the products or from some other process, to provide part of the energy used in the separation. While their objectives remain the same, and their results may be comparable, recovery brings with it an interaction between the column being controlled and the source of heat.

For these and other reasons, it will be found more effective to improve a column's efficiency first, and then consider recovering heat.

1. Floating-Pressure Control

Most volatile mixtures are easier to separate as their pressure is reduced, because of a favorable change in relative volatility with pressure and temperature, as described in Example 8.4. Advantage may be taken of this property during hours when the heat-removal capacity of the condenser is more than sufficient to keep the column pressure under control.

Most columns are pressure-controlled by manipulating the condenser in some fashion. The coolant or the condensate may be throttled as shown in Fig. 1.5, or the vapor may be throttled and/or bypassed around the condenser. A less satisfactory method involves pressurizing the condenser with a non-condensible gas, which is vented should the pressure rise too high.

In every case, the application of controls raises the column pressure higher than it would be if the condenser were left unrestricted. The cost of increasing the irreversibility of this heat-transfer process is the use of more energy to make the separation than would be required if the column were operated at minimum pressure.

It is not possible to predict with certitude exactly what the minimum pressure might be at any point in time. However it is relatively easy to *control* a column at the minimum pressure simply by driving the pressure control valve to an extreme position. Care must be exercised not to forego pressure control altogether, however. In the absence of control, the pressure in some columns has been observed to oscillate due to interaction between the condenser, overhead accumulator, and reflux flow. This oscillation can be eliminated with tight pressure control.

Another problem is the severe upset that can develop when a fan condenser is exposed to a sudden rainstorm. The efficiency of the condenser increases markedly when it is wet, and the resulting increase in heat-transfer rate can cause pressure to fall rapidly. In the absence of control, the falling pressure will reduce the boiling point of the liquid throughout the column, converting sensible heat to latent heat. The resulting temporary increase in the rate of boiling has been known to flood a column.

These problems may both be avoided by controlling pressure in the short term, while minimizing its set point in the long term. Figure 8.9 shows how this is accomplished. A conventional pressure controller operates the condenser throttling and/or bypass valve in the usual manner. The controller output is then sent to a valve-position controller, which slowly manipulates the pressure set point until the valve(s) are driven to a specified position near the limit. For example, a condenser bypass valve would be driven gradually to a 10% open position by adjusting the pressure set point. This margin is allowed so that the pressure controller is able to respond to small upsets in either direction; large upsets are likely in the direction of decreasing pressure, in that weather conditions are likely to deteriorate more rapidly than they improve. A sudden rainstorm will cause the bypass valve to open to keep pressure from falling. Then the pressure set point will gradually be reduced until the valve has returned to its set position. The response of valve position, and hence pressure, to a step change in cooling availability is exponential, having a time constant equal to that of the integral mode of the valve-position controller.

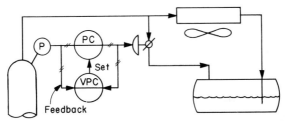

Fig. 8.9.   The valve-position controller acts to keep the vapor bypass valve nearly closed in the steady state.

The improvement in relative volatility brought about by the pressure reduction will appear in improved quality of one or both products, unless quality control is enforced. If it is enforced, then the results of reduced pressure appears as a savings in energy. There are some problems associated with bringing this about, however.

Quality-control loops tend to be slow. If a change in column pressure is to avoid causing undue variations in product quality, the integral time of the valve-position controller must be an order of magnitude longer than that of the composition controllers.

The speed of response of the entire system may be increased through the use of feedforward control. It is possible to calculate the relationship between

8. Distillation

Table 8.2

**Heat-Input-to-Feed Ratios for
Pentane Separation**

Temperature (°F)	Pressure (psia)	H_v (Btu/lb)	α	β	$H_v\beta$
140	38	135	1.25	0.600	81.0
120	28	139	1.28	0.521	72.4
100	20	143	1.32	0.440	62.9
80	14	146	1.35	0.392	57.2

the heat-input-to-feed ratio, and column pressure, to apply feedforward control. Equation (8.14) stated the heat input required to reach a given separation. For a constant separation, then, Q^*/F varies directly with $H_v\beta$. Table 8.2 lists $H_v\beta$ versus pressure and temperature for n-pentane–isopentane splitter having 70 theoretical trays. Equation (8.10) was used to make the calculations.

When $H_v\beta$ is plotted against pressure as in Fig. 8.10, the points fall in a straight line. (The small amount of scatter appearing is due to the values of α and H_v being available in only three significant figures.) The line can be

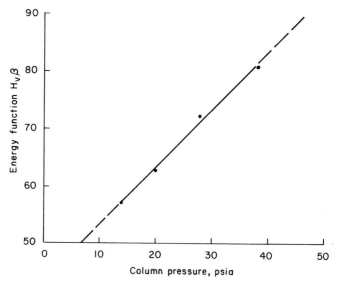

Fig. 8.10. The energy required to separate a unit of feed can be expected to vary linearly with column pressure.

represented by an equation of the form

$$H_v\beta = k_1(p + k_2) \qquad (8.24)$$

This particular form was chosen to separate parameters peculiar to the column from those peculiar to the components. Terms H_v and α are characteristic of the key components, and their variation with pressure bears on both k_1 and k_2. However the number of trays and their efficiency do not uniquely vary with pressure, and therefore affect only k_1. For the n-pentane–isopentane separation, k_2 is 43.7 psi; k_1 must be left adjustable to allow for variations in column parameters.

Figure 8.11 describes a feedforward system for control of bottom composition, using column pressure and feed rate to manipulate heat input. This system actually contains a positive feedback loop in that column pressure tends to increase with heat input. But the regulation of the pressure controller of Fig. 8.9 can overcome any stability problems that may appear. Pressure can only change as fast as the valve-position controller will allow. So proper selection of the integral time constant will provide stabilization, without the need for additional dynamic compensation.

The energy savings attributable to increased cooling availability may be estimated through the use of Table 8.2. Consider that an air-cooled condenser removes heat to the surroundings at 90°F under design conditions. As ambient temperature falls at night, or with changes in weather and season, the pressure could be maintained at its maximum of 38 psia by bypassing the condenser or turning off fans, or both. This is the conventional operating practice. If instead, floating pressure control is applied, reflux temperature and column pressure will follow changes in ambient temperature. At lower pressures, less energy is required to make the separation, which reduces the temperature difference across the condenser, allowing a closer approach to

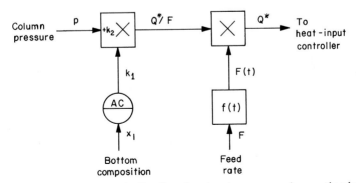

Fig. 8.11. This system automatically adjusts heat input as pressure changes, thereby saving energy without upsetting product quality.

Table 8.3

Heat Input versus Ambient Temperature for
Air-Cooled Columns

Reflux temperature (°F)	Heat input (% of base)	Ambient temperature (°F)
140	100	90
120	89.4	75.3
100	77.7	61.2
80	70.6	44.7

ambient. Following this line of reasoning, Table 8.3 was prepared, listing the ambient temperatures that correspond to the column pressures of Table 8.2.

The average reduction in heat input is 0.65%/°F reduction in ambient temperature. These savings compare favorably with the 0.72%/°F predicted for a depropanizer in Ref. (2), and 0.68%/°F for a deisobutanizer in Ref. (1, p. 214). Actual operating experiences generally confirm these estimates.

Distillation experts express concern that reduction in pressure will bring a column closer to flooding. In actual practice, however, the boilup required to maintain a constant separation decreases more with pressure than does the calculated flood point. The pentane splitter operates farther from its flood point at 14 psia than at 38 psia. The difference is even greater for lighter hydrocarbons. However feedforward adjustment of boilup from column pressure is necessary to avoid a temporary flooding condition in columns separating butanes and heavier components. Otherwise pressure could fall fast enough to cause flooding before a commensurate reduction in heat input could be made. Columns separating propane and lighter components overhead do not need this protection, since the heat input at which they flood increases as pressure is reduced, as described in Ref. (4).

Some columns are equipped with partial condensers, with pressure being controlled by continuously venting noncondensible components. A reduction in condenser temperature under conditions of constant pressure will raise the vapor pressure of the distillate product. To avoid this, many processors control reflux temperature by bypassing or throttling the condenser.

Vapor pressure can be controlled without bypassing or throttling if column pressure is set in accordance with condensate temperature as shown in Fig. 8.12. The relationship between pressure set point and measured temperature should be adjusted to follow the desired product vapor-pressure curve. Over the range of temperatures normally encountered, the relationship is essentially linear.

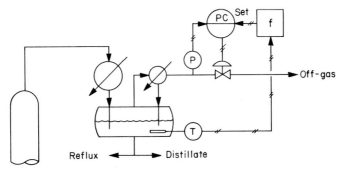

Fig. 8.12. Column pressure should be set by condensate temperature, to control its vapor pressure.

The same sort of relationship must be used if temperature control is applied to any column whose pressure is variable. The temperature measurement may be compensated for variations in pressure from base (b) conditions:

$$T_b = T + \frac{\partial T}{\partial p}\bigg|_{x^*} (p_b - p) \tag{8.25}$$

where the partial derivative represents the slope of the vapor-pressure curve for the desired product composition x^*. Measured variables are temperature T and pressure p, and T_b is the temperature which would be measured if the pressure were controlled at p_b.

Another technique involves the measurement of differential vapor pressure (5). A temperature bulb partially filled with a sample of the product is connected to one side of a differential-pressure transmitter. The bulb is inserted at an appropriate point in the column, with the other side of the transmitter connected to the column at the same elevation. The transmitter then reports a difference in vapor pressure between the liquid in the column and the liquid in the bulb. It has been found particularly effective in controlling the proof of distilled alcohol products.

Producer Pairity Control on N_2 Dist Column Unless Carbor Used

2. Heat-Pump Systems

Heat pumps are being used for separation between close-boiling components such as propane and propylene, and normal and isobutane. Overhead vapor is compressed, raising its pressure and temperature to the point where it can transfer heat to the bottom product. It is then condensed by reboiling the bottom product, with the condensate returned as reflux, as shown in Fig. 8.13.

The relationship between work and heat transfer was given in Chapter 5:

$$-W = Q_c(T_c - T_v)/\eta T_v \tag{8.26}$$

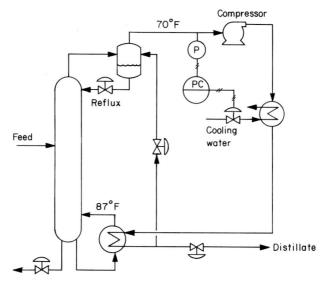

Fig. 8.13. Overhead vapor is compressed enough to raise its boiling point above that of the bottom product; the net energy input of the compressor is removed by cooling.

where Q_c is the heat removed from the vapor condensing at T_c, T_v is the overhead vapor temperature in absolute units, and η is the efficiency of the cycle. The heat pump is most attractive where $T_c - T_v$ is small, and Q_c is large, as in the separations mentioned above.

As an example, consider the separation between propane and propylene. If T_c can be held at 70°F, the pressure in the column will be about 152 psia, the vapor pressure of pure propylene. At a maximum vapor load, a typical perforated tray will develop about 2 in. of water differential pressure. A column with 110 trays would then develop perhaps 220 in. or 8 psi differential. Propane at the bottom of the column would then boil at 89°F at 160 psia. Reference (6) gives the optimum temperature difference across the reboiler tubes as about 16°F for smooth tubing and 8°F for high-flux tubing. The temperature difference $T_c - T_v$ would then be 35 and 27°F, respectively, for the two types of tubing. Using an efficiency factor of 70%,

$$-W/Q_c = 35/0.7(530) = 0.094$$

for smooth tubing, and

$$-W/Q_c = 27/0.7(530) = 0.073$$

for high-flux tubing. When corrected for the 3-to-1 relationship between fuel heat content and work content, the amount of fuel energy required to transfer a unit of heat is only 0.28 and 0.22 for the two cases.

If a heat pump were applied to the propane–propylene column of Fig. 8.3, the work required to separate 1 lb-mol of feed with high-flux tubing would be

$$\frac{-W}{F} = \frac{Q}{F}\left(\frac{-W}{Q}\right) = 24{,}600(0.073) = 1790 \quad \text{Btu/lb-mol}$$

Using a heat-sink temperature of 70°F (cooling water) the entropy increase in the surroundings experienced using this amount of work is

$$\Delta S = \frac{-W}{T_0} = \frac{1790}{460 + 70} = 3.39 \quad \text{Btu/lb-mol-°F}$$

This increase in entropy is only 2.7 times the entropy reduction in the separated components, compared to the 7.7 factor using low-pressure steam as a heat source. Heat pumps are thereby demonstrated to conserve entropy.

Bear in mind, however, that columns separating these components may not typically be supplied with fuel energy but often with waste heat from somewhere in the plant. This could tip the balance away from using a heat pump.

External cooling is necessary to remove the energy introduced by the compressor, and to adjust for differences in sensible heat between feed and products, and heat losses. A pressure controller is shown manipulating cooling just as in a conventional column, although the size of the cooler is a small fraction of the size of the condenser-reboiler. Boilup rate is set by the speed of the compressor.

Reference (6) lists some further opportunities for improved efficiency. A small second stage of compression can be added to pump heat to the external cooler, reducing the pressure of the column and the condenser-reboiler. This can lower fabrication costs and improve relative volatility, which in turn can result in fewer trays or reduced energy requirements. Through the use of trays with lower operating pressure drop, $T_c - T_v$ can be reduced further.

Finally, floating pressure control is as applicable to the heat-pumped column as to any other. A reduction in coolant temperature will allow a commensurate reduction in column temperatures, even if a second stage of compression is used. But because only a small fraction of the heat flowing through the column is removed by the condenser, the response of pressure to disturbances in the cooling medium will be an order of magnitude slower than for a conventional column.

3. Compressor–Expander Systems

Cryogenic processing needs to be energy-efficient to be viable at all. As a consequence, considerable effort has gone into building expansion turbines for such gases as nitrogen in air-separation units, and methane in condensate recovery plants. Figure 8.14 describes a plant using an expander to help recover light hydrocarbons (condensate) from natural gas in a stripping column.

Natural gas at 1000 psig and ambient temperature is cooled counter-currently against stripped methane to about $-100°F$, causing heavier components to condense. These components are then fed to the column under level control. The vapor phase is expanded through either an engine or a valve. The engine recovers work for recompressing the stripped gas, and in so doing condenses about 30% of its vapors, according to Ref. (7). Some lique-faction also takes place across the valve due to the Joule–Thompson effect, but because no energy is extracted only about 10% of the discharge is liquid. For maximum performance, the valve should be closed, yet it is useful in controlling pressure by adjusting the energy balance in the process.

If pressure must be controlled, a valve-position controller should index the set point to keep the bypass valve in a nearly closed position. The pressure set point will then vary with throughput, and feed gas temperature and pressure.

Because of the limited efficiencies of both the expander and its attached compressor, the product gas can only be restored to a fraction of its supply pressure. A motor-driven compressor is then needed to restore the product

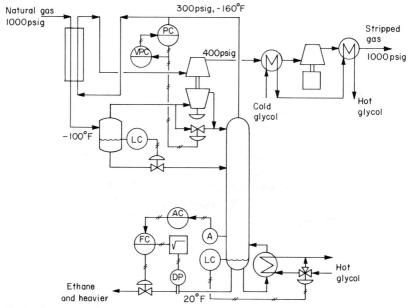

Fig. 8.14.   Pressure in the stripping column can be controlled by the expander bypass valve, but at the cost of losing both reflux and shaft work.

to line pressure. Waste heat from recompression can be used for reboiling as shown; a glycol water solution is the heat-transfer medium because water would freeze at the reboiler temperatures encountered.

Because the bottom product is only a small fraction of the feed, it can be manipulated quite accurately to control its methane content. This places base level under control of boilup to balance the variable amount of liquid flowing down to column.

D. ENERGY RECOVERY TECHNIQUES

The last example smacks of energy recovery, and possibly might have been included under this heading. Yet recovery systems have the unusual property of introducing potential dynamic instability. If energy is recovered from a product to a feed, a feedback loop is formed, in some cases negative, in others positive. Double-effect systems are characterized by parallel flows of mass and energy, requiring balancing for quality control. And when energy is recovered from one process for use in another, control loop interactions can arise. Each of these situations is examined with an eye to maximize efficiency without sacrificing stability.

1. Regenerative Heat Recovery

The most common example of regenerative heat recovery is the use of bottom product to preheat the feed to a column. Occasionally the bottom product will come from a downstream column, but usually it is from the same column the feed enters. Often a steam-heated exchanger will be used for additional preheat.

The stability problem is caused by feedback from bottom heat, which changes the liquid–vapor distribution in the column. For example, a rising base level will cause the level controller to increase bottom-product withdrawal. This increases heat transfer to the feed, thereby reducing the liquid flowing into the column base. The sense of the feedback loop is negative in that it will lower base level just as increasing product withdrawal does. However the feedback is delayed by the response of the heat exchanger and the trays in the lower section of the column. A deviation in base level may already be corrected by bottom-product flow before the effect of feed preheat arrives—then an overcorrection results.

Whether this loop is sufficient to destabilize base level depends on the characteristics of the reboiler. With a thermosyphon reboiler, a level is maintained in the *base* of the column, which tends to have more capacity than the *reboiler*. However, with a kettle reboiler, the column base is empty, and level control is confined to a small volume at the end of the reboiler. Because of the small capacity of the chamber where level is measured, kettle reboilers are much more difficult to control than other types, and are susceptible to destabilization through the preheater.

Detuning the level controller does not solve the problem—it only causes wider and slower fluctuations. Stability can be provided by tightly controlling feed temperature. Because the period of the temperature control loop is much shorter than the feedback loop through the column, the slow level cycles are eliminated. However in order to control feed temperature, some of the bottom heat must be rejected.

Figure 8.15 shows how to maximize heat recovery without sacrificing stability. The set point of the feed temperature controller is adjusted by a valve-position controller to keep the valve nearly full open to the preheater in the steady state. This system functions identically to the floating pressure control system. A set point of about 90% is recommended for the valve-position controller to allow an adequate margin for the temperature controller to function properly.

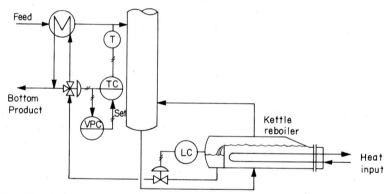

Fig. 8.15. Regenerative preheating requires tight temperature control to stabilize the level loop; heat recovery is maximized by the valve-position controller.

An alternate installation uses two preheaters: the first recovering all available heat from the bottom product; the second adding preheat from steam, under temperature control. The same system can be used for the steam-heated exchanger. The position of the steam valve can be held at 10% open in the steady state by manipulating the set point of the temperature controller. In this system, heat recovery is already maximized—the valve-position controller functions to *minimize* steam usage.

2. Double-Effect Operation

Distillation columns can be cascaded like evaporators, with heat and process streams flowing cocurrently, countercurrently, or with process flow split between parallel units. The series cocurrent configuration shown in Fig. 8.16 is probably the most common. It is used to separate three products from a multicomponent feed.

The boiling point of the intermediate product at the bottom of the second column tends to be higher than the dew point of the overhead vapor from the first. In order to transfer heat from the first column to the second, then, the first must operate at a higher pressure. But the pressure in the first column *cannot* be controlled by manipulating the valve feeding the second. If the pressure in the second column is constant, then the pressure in the first as determined by its condenser temperature will therefore vary with heat load and intermediate-product composition. If the valve feeding column 2 is manipulated to hold the pressure in column 1 constant, composition in the condenser-reboiler must then change with heat load.

Ideally, the pressure in column 1 should not be controlled at all, but should float on the condenser-reboiler. Pressure in column 2 may be controlled, but energy will be saved if it also is allowed to float on its condenser. A floating pressure control system as in Fig. 8.9 should be applied to column 2, with feedforward control from column 1 pressure applied to its reboiler, as in Fig. 8.11.

A second control problem is the inability to manipulate boilup independently in the second column. An auxiliary reboiler is indicated in Fig. 8.16,

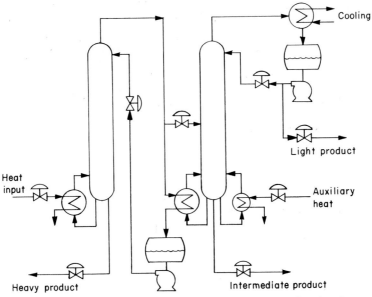

Fig. 8.16.   All of the energy used to boil-up the first column is transferred to the second.

but its use defeats the purpose of double-effect operation. Column 1 boilup can be adjusted to control the intermediate key in the heavy product, with feed to the second column manipulated to control the heavy key in the intermediate product. The material balance in column 2 may control either the light key in the intermediate product or the intermediate key in the light product, but not both at the same time. If feed composition is uniform, control of one will regulate the other. If it is variable, a decision has to be made regarding which of the two product specifications must be sacrificed. If one product is more valuable, its quality must be controlled at the cost of losses in the other product, when separation is inadequate. In effect, the series-cascaded columns are controlled like a single sidestream column.

Parallel cascading, as shown in Fig. 8.17, introduces a balancing problem as well as a pressure-control problem. The first-effect column must operate at higher pressure than the second effect, as with the series cascade arrangement, and may be allowed to float on the condenser-reboiler. If the columns have the same number of trays, the Q/F ratio in column 1 must be less than that in column 2 because of the reduced relative volatility at the higher pressure. Since heat flow Q is the same through both, their individual feed rates must differ as shown in the figure.

Energy usage will be minimized if all compositions meet specifications. In absence of hard specifications on one product, streams from parallel columns should still have identical compositions if they are going to the same place. If the characteristics of the columns are similar, there should be no difficulty in matching separation factors, simply by maintaining a constant

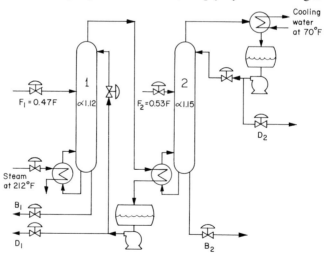

Fig. 8.17. In a parallel double-effect system, the feed rates must be properly balanced to achieve identical separation from the same heat flow.

ratio of the two feed streams. Variations in feed composition, feed rate, and pressure should affect both equally, so that readjustment of the feed ratio would seldom be required. The heat input of column 1 could be used to control its bottom-product composition, and that of column 2 should simply follow it.

If the same heat source and cooling are applied to the columns in Fig. 8.17 as to the single column of Fig. 8.3, the heat requirement and entropy gain are both reduced to 53% of that of the single column. Thus double-effect distillation does, like evaporation, reduce heat input requirements almost in half. The 53% factor does not include heat losses, however, which were included in the discussion on heat economy of evaporators.

Multiple-effect distillation conserves entropy by making use of the available work in the heat source. If water at 140°F were the heat source, only a single propane–propylene column could be reboiled because the water would not have enough available work. Perhaps the greatest contribution of multiple-effect operations is to make use of the avilable work that exists in heating media, and to draw attention to its importance in designing a process.

3.   Coupling Unrelated Columns

Reference (8) describes a procedure for arranging two-product towers to separate a multicomponent feedstock at minimum cost. The program includes energy integration, where the condenser of one column is the reboiler of another. This goes a step beyond multiple-effect distillation in that the coupled columns may not be in either a direct serial or parallel flow pattern. Furthermore, heat duty may be split—the overhead vapor from one column could be used to reboil two other columns.

When heat integration is as intensive and varied as this, the sources and sinks may be considered as completely unrelated, regardless of whether they actually are. The point remains that they have the capability of disturbing one another, and this interaction must be removed, regardless of whether it includes feedback loops within the process.

A case in point is the use of heat rejected by an atmospheric crude-oil column to reboil the debutanizer in Fig. 8.18. The debutanizer may be refining the light naphtha cut from the crude column, or processing a completely different stream. Disturbances in the crude column can upset the vapor loading in the debutanizer, and changes in the heat demand of the debutanizer can affect reflux in the crude column.

Figure 8.18 shows two control systems designed to eliminate this interaction. Differential pressure measured across the debutanizer indicates its vapor flow, similar to pressure drop across an orifice. Perforated trays probably give the most consistent relationship between vapor flow and

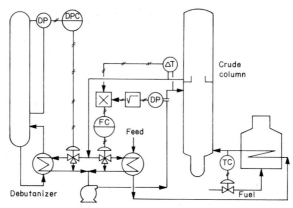

Fig. 8.18.   Heat removed from a crude column to adjust internal reflux may be recovered to reboil a debutanizer.

differential pressure, with some influence from liquid flow. Valve trays tend to regulate differential pressure, so that it changes little with vapor flow over the range where the valves are partly open. However, many valve-tray columns are operated at rates where the valves are fully open, and differential pressure is again proportional to vapor flow.

The differential-pressure transmitter must be correctly installed to provide an accurate indication of boilup. Vapors will tend to condense in the lines leading to the transmitter, causing measurement errors. For best results, the low-pressure connection should be as close to the process as possible; connecting to the vapor line entering the condenser as shown in Fig. 8.19 will minimize the physical elevation of the transmitter. The line connecting the

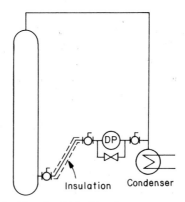

Fig. 8.19.   The connecting line should be $\frac{3}{4}$-in. pipe or larger, should drain freely, and be insulated to minimize refluxing.

base of the column to the higher-pressure tap should be of $\frac{3}{4}$-in. pipe or larger, and drain freely. Insulation will minimize refluxing in this line, improving the quality of the signal. Stop valves should be of the ball type, which features a line-size unrestricted opening. A bypass valve allows adjusting the zero calibration of the transmitter while the column is operating. Purging the connections with inert gas is not recommended, as it raises column pressure, reducing relative volatility, and carries away product when vented.

Figure 8.18 also shows a heat-flow controller to isolate the crude column from changes in heat flow to the debutanizer. Its input is calculated from a measurement of recirculation flow rate and temperature difference. Whatever heat is not needed by the debutanizer is sent to the feed preheater to maintain a constant rate of heat removal. A temperature controller at the outlet of the fired heater adjusts fuel flow as necessary to compensate for variations in preheater outlet temperature. If variations are severe, feedforward control can be added to manipulate fuel flow from a measurement of preheater outlet temperature.

While this discussion centered on transferring of heat from one column to another, the same approach applies to the use of any waste heat source for reboiling a column. In olefins plants, it is common to use hot water from quenching the products from the cracking furnaces to reboil the propane–propylene column. Because this source of energy is free, its use may result in a lower-cost installation than a heat-pumped column. But waste heat tends to be variable in temperature and flow. A differential-pressure control loop as shown on the debutanizer in Fig. 8.18 has been very successful in regulating the flow of waste heat to a propane–propylene column.

REFERENCES

1. Shinskey, F. G., "Distillation Control: for Productivity and Energy Conservation." McGraw-Hill, New York, 1977.
2. Fauth, C. J., and F. G. Shinskey, Advanced control of distillation columns, *Chem. Eng. Progr.* (June 1975).
3. Shinskey, F. G., "Process-Control Systems." McGraw-Hill, New York, 1967.
4. Shinskey, F. G., Energy-conserving control for distillation units, *Chem. Eng. Progr.* (May 1976).
5. The Foxboro Company, Differential vapor pressure cell transmitter, Model 13VA, *Technical Information Sheet* 37–91A, Foxboro, Massachusetts (April 1965).
6. Wolf, C. W., D. W. Weiler, and E. G. Ragi, Energy costs prompt improved distillation, *Oil Gas J.* (Sept. 1, 1975).
7. Houghton, J., and J. D. McLay, Turboexpanders aid condensate recovery, *Oil Gas J.* (Mar. 5, 1973).
8. Rathore, R. N. S., K. A. VanWorme, and G. J. Powers, Synthesis of distillation systems with energy integration, *AIChE J.* (Sept. 1974).

Part

IV

ENERGY SYSTEMS IN BUILDINGS

Chapter

9

Heating, Ventilating, and Air-Conditioning

Much of our fuel is consumed in attempting to provide a comfortable environment for the occupants of homes and buildings. The word "attempting" is used intentionally in that the conditions within which one must live and work are a frequent source of complaint for most of us. In many cases, the complaints are justifiable in that the heating, ventilating, and air-conditioning system—hereafter abbreviated HVAC system—is deficient in one or more respects. But comfort is also a subjective issue—conditions that are comfortable to one individual may not be to others.

Energy conservation in HVAC systems must be approached from three directions. The equipment must be designed and configured for maximum efficiency: it must be controlled for minimum energy expenditure consistent with its objectives, but of equal importance is the proper conditioning of the occupants. As the seasons change, the metabolism of warm-blooded creatures slowly adjusts. An unseasonably cool day in the summer can cause more discomfort than a colder day in midwinter, because of the relative condition of blood viscosity and other parameters which affect metabolic heat transfer.

In attempting to create uniform indoor conditions year round, the HVAC engineer is doing a disservice to the occupants. A person living in 60–65-°F surroundings will be less affected upon exposure to 0–25-°F outdoor temperatures than one living in a 70–75-°F environment. He will also be less susceptible to illness and use less fuel. Because energy usage depends heavily on behavior patterns, this chapter opens with a discussion on comfort—what it means and what it costs.

A. DEFINING A COMFORTABLE ENVIRONMENT

In other applications described earlier, specifications of control set points were generally accepted, and systems were developed to maintain variables at those set points with the least flow of energy. In HVAC systems, however, the controlled variables of temperature, air velocity, and humidity combine to produce comfort conditions that are largely subjective. The first step in defining comfortable conditions begins with an examination of the metabolic functions of the human body.

1. Metabolic Functions

The human body is equipped with a very sensitive and reliable system to control its core temperature at 98.6°F. Heat is generated within the body by metabolism—the oxidation of food—and released by heat transfer. The metabolism of humans at rest, referred to as "basal metabolism," lies between 300 and 400 Btu/h for the average man. It is slightly less for women and elderly persons, and somewhat greater per unit body surface area for children. Basal metabolism is not appreciably affected by ambient temperature over the normal comfort range of 60–85°F. However exposure to lower temperatures will promote muscle tension increasing metabolism, and higher temperatures will bring about inactivity and fatigue, decreasing it.

The level of work activity has a pronounced effect on heat generation. Men absorbed in vigorous physical labor can generate as much as 1300 Btu/h of heat. While such an individual can remain comfortable in ambient temperatures below 30°F, his production will begin to decline when the ambient exceeds 65°F. Although most HVAC systems are applied to residences, schools, hospitals, offices, etc., where occupants are essentially at rest, this factor needs to be considered for other working environments.

In the ambient range up to 85°F, body core temperature of the adult at rest is controlled by manipulating heat-transfer through a combination of radiation and convection. The relationship between heat-transfer rate by these mechanisms and ambient temperature is essentially linear, falling from perhaps 400 Btu/h at 55°F to zero at 99°F. At temperatures below 85°F, perhaps 100 Btu/h of heat is lost by evaporation, principally in exhalation.

Above this point, evaporative losses begin to increase sharply, eventually carrying the full heat load at 99°F. For persons at work, the rate of evaporation begins to increase at a lower temperature, in proportion to the metabolic rate.

2. Effective Temperature

The differences between these heat-transfer mechanisms bear heavily on the condition of the air which feels comfortable to an individual. In a cool environment, blood vessels are constricted, reducing the temperature of the skin. This action compensates for the low air temperature, keeping the heat-transfer rate close to the metabolic rate. Clothing reduces the skin area exposed to heat transfer.

If heat is being applied to a room to maintain a controlled temperature, increasing air velocity will improve heat transfer from the heating coils to the room air, but also from the body to the air. Moving air makes a person feel cool by increasing the rate of heat transfer from the skin. Comfort then requires higher air temperatures when air velocity is high. Furthermore, moving air also promotes heat losses through outside walls and windows. To conserve energy during the heating season, airflow should be minimized.

A term called "effective temperature" was coined by the American Society of Heating, Refrigeration, and Air-Conditioning Engineers (1) to describe a particular degree of comfort, related to wet-bulb and dry-bulb temperatures, and air velocity. In still air (15–25 ft/min) at its dew point, the effective temperature is equal to the dry-bulb temperature. Lower humidities and higher velocities will reduce the effective temperature. To determine the conditions that constitute a particular effective temperature, people were asked to compare the comfort of a test room against that of a room controlled at 100% humidity and having negligible air movement. When they felt equally comfortable in each, the conditions in the test room were considered to produce the same effective temperature as in the controlled room.

The effective temperatures recommended by the ASHRAE in 1959 (2) were a low of 66°F in winter and a high of 73°F in summer. The winter figure is now considered higher than necessary and desirable for conservation. A value of 62°F is more in line with current goals, and even lower temperatures may be requested in the future. In countries where energy has historically been scarce and costly, lower temperatures are common.

As the effective temperature in homes, schools, and workplaces is moved closer to outdoor temperature, there will be less shock to the metabolic system as people travel from place to place. Additionally, clothing can be worn that will be equally comfortable at home, outdoors, and at work. While conserving fuel, these conditions are, in the experience of the author, also more healthful.

During the cooling season, an effective temperature of about 73°F seems most conducive to a high level of production in work areas (2). While people can wear heavier clothing against a decreasing effective temperature, their response to an increasing temperature is more limited, and the temperature at which productivity begins to fall is not likely to change. However an effective temperature of 73°F is still much higher than that maintained in many modern air-conditioned buildings. A controlled dry-bulb temperature of 78°F, for example, can produce an effective temperature of only 68.5°F in a typical air-conditioned space where the relative humidity is maintained at 50%, and the air velocity is 300 ft/min.

A very important step in conserving energy is to enforce an effective temperature of 62°F while heat is being applied, and 73°F while cooling is used. When neither are required, the effective temperature can be free to float between these limits. The next consideration is to select sets of conditions that produce the specified effective temperatures with a minimum of energy.

3.   The Heating Season

Table 9.1 compares the enthalpy and dry-bulb temperature against effective temperature for still air (15–25 ft/min) at 30% R.H. Air enthalpy is an indication of how much heat must be supplied to elevate the outside air to the stated conditions. Outside air in the wintertime, with a 30°F dry-bulb temperature and a relative humidity of about 50%, will have an enthalpy of 9 Btu/lb. To heat this air to an effective temperature of 66°F versus 62°F requires 21% more energy. Furthermore, convective and conductive heat losses from a building at 72°F to the outdoors at 30°F are 15% higher than from the same building at 66.5°F. Therefore, the use of 62°F effective tem-

Table 9.1

**Enthalpy and Dry-Bulb Temperature versus Effective
Temperature in Still Air at 30% Relative Humidity**

Effective temperature (°F)	Dry-bulb temperature (°F)	Enthalpy (Btu/lb)
66	72.0	22.6
64	69.0	21.4
62	66.5	20.2
60	63.5	19.1
58	61.0	18.1

perature rather than 66°F can save between 15 and 21% in fuel consumption, depending on the relative magnitude of convection and conduction versus infiltration losses.

Table 9.2 illustrates the influence air velocity has on maintaining an effective temperature of 62°F. Approximately 32% more energy is required to heat outside air at 30°F and 50% R.H. to a 62°F effective temperature at 500 ft/min as would be required in still air. Furthermore, heat losses by convection and conduction are 22% higher at 500 ft/min due to the higher dry-bulb temperature, without taking into account the effect of the velocity on heat transfer. All things considered, as much as 30–35% more energy may be required to maintain 62°F effective temperature at 500 ft/min than in still air.

Table 9.2

Enthalpy and Dry-Bulb Temperature Required for
62°F Effective Temperature at
30% Relative Humidity

Velocity (ft/min)	Dry-bulb temperature (°F)	Enthalpy (Btu/lb)
20	66.5	20.2
100	69	21.3
300	72.5	22.8
500	74.5	23.8

While minimum effective temperature and air velocity both result in energy savings, the role of humidity is not as clear. During the heating season, inside air tends to be quite dry because of the low absolute humidity of the outside air. Outside air at 30°F dry-bulb temperature and 50% R.H., will decrease to only 15% R.H. when heated to 67.5°F, which would be required for 62°F effective temperature in still air. This is a typical value for humidity inside heated buildings in northern U.S. winters.

Lower temperatures become more comfortable as humidity is increased, as indicated in Table 9.3. While this would appear to save heat losses through conduction and convection, the higher enthalpy of the air augments infiltration losses. Compare heat-transfer losses to infiltration losses for inside air at 50% and 10% R.H., with outside air at 30°F and 50% R.H. Inside air at 50% R.H. has 58% more heat loss through infiltration, and only 8.6% less heat loss through conduction and convection.

If the rate of air infiltration through a building can be kept very low, then a high humidity level can be maintained, with a reasonable expenditure of

Table 9.3

**Enthalpy and Dry-Bulb Temperature Required for
62°F Effective Temperature in Still Air**

Relative humidity (%)	Dry-bulb temperature (°F)	Enthalpy (Btu/lb)
100	62	27.8
70	64	25.1
50	65	22.6
30	66.5	20.2
10	68	17.6

energy. This could be particularly important in a building with considerable glass or metal surfaces, whose heat losses are great. But where rapid air changes are necessary, as in a chemical laboratory or a gymnasium, a high humidity cannot be economically maintained.

Extremely low humidities create problems with the generation of static electricity, deformation of wooden structures, and irritation of skin and respiratory tissues. Overnight reduction in dry-bulb temperature can moderate these effects to a certain extent. Yet it appears to be desirable to maintain a minimum relative humidity of about 30% during the heating season. Again, the optimum value will vary with the relative rates of heat loss by convection and conduction, compared to infiltration.

The rate of ventilation maintained in a building has a substantial impact on heating requirements, and also bears directly on the optimum humidity. ASHRAE recommends a minimum of 5 ft^3/min fresh air per person for non-smoking areas and 25 ft^3/min for smoking areas. The General Services Administration (3) reports that the oxygen requirements can be met with only 1–2 ft^3/min per person. Ventilation rates can be reduced toward this minimum if filtration and adsorption are provided to remove smoke and odors.

4. The Cooling Season

Many air-conditioned buildings are cooled more than necessary or desirable, both from an energy savings point of view and for the health of their occupants. Substantial savings can accrue by raising thermostats to attain an effective temperature of 73°F, as indicated in Table 9.4. To cool outside air having a 90-°F dry-bulb and 60% R.H. to an effective temperature of 67°F requires more than twice the heat removal than required to cool it to an effective temperature (E.T.) of 73°F. Furthermore, heat gain by conduction and convection is 80% greater at the lower temperature. Much of the

Table 9.4

Enthalpy and Dry-Bulb Temperature versus Effective
Temperature in Air at 300 ft/min and 60% R.H.

Effective temperature (°F)	Dry-bulb temperature (°F)	Enthalpy (Btu/lb)
67	75.5	30.3
69	77.5	31.9
71	80.0	33.1
73	82.0	34.8
75	84.0	36.6

cooling load is required to remove basal metabolism, heat from lighting and machinery, and solar radiation, which are not particularly related to E.T. Therefore the energy savings attributable to an increase in E.T. from 67 to 73°F will be less than 100%, depending on occupancy and other factors, but is nonetheless considerable.

The effect of air velocity on comfort appears in Table 9.5. Over three times as much energy must be removed from outside air at 90°F and 50% R.H. to reach an E.T. of 73°F in still air as in air moving at 500 ft/min. Although the temperature difference between inside and outside is doubled, in still air heat transfer through walls and windows may actually be lower.

Table 9.5

Enthalpy and Dry-Bulb Temperature Required for
73°F Effective Temperature at 60% R.H.

Velocity (ft/min)	Dry-bulb temperature (°F)	Enthalpy (Btu/lb)
20	77.5	31.5
100	79.0	33.1
300	82.0	34.8
500	84.0	36.2
700	85.0	37.6

Humidity, although not specifically controlled in an air-conditioned environment, is important to comfort. Table 9.6 gives various conditions producing an E.T. of 73°F, noting the dew point of the air. Here again, humidity and dry-bulb temperature act in opposition on energy requirements. High humidity levels require little heat removal from outside air, yet require comparatively higher reduction in dry-bulb temperature. The two effects may virtually cancel each other.

Table 9.6

Enthalpy and Dry-Bulb Temperature Required for 73°F Effective Temperature in
Air at 300 ft/min

Relative humidity (%)	Dry-bulb temperature (°F)	Dew-point temperature (°F)	Enthalpy (Btu/lb)
80	79.5	73	38.7
70	81.0	70	36.6
60	82.0	66	34.8
50	83.0	62	33.1
40	84.0	58	31.5

Air dew point is established by the temperature of the cooling surface to which it is exposed. Where chilled water at 40°F is supplied to the air-conditioning unit, a dew-point temperature in the region of 55°F can be expected. Under these conditions, a dry-bulb temperature above 84°F would be tolerable. However, this does not represent an efficient mode of operation for the refrigeration unit. Less work would be required to provide chilled water at 55°F, yielding a dew point of perhaps 70°F. When combined with the closer approach to outside-air enthalpy, it should more than offset the modest reduction needed in dry-bulb temperature.

B. HVAC SYSTEMS

There is a wide variety of systems in common use, each having its own advantages and control problems. Rather than attempt to cover the entire spectrum, the following describes the general features most often encountered, and how these units are most efficiently controlled. New systems incorporating solar technology are discussed later in the chapter.

1. Reheat Systems

A multizone HVAC system with reheat is shown in Fig. 9.1. A centrally located fan draws a mixture of return air and fresh air through chilled-water coils, distributing it to several ducts containing steam coils. Each duct serves a particular zone where temperature is controlled by a "master" unit. The master sets the temperature of the "submaster" controller, whose sensor is located in the duct, downstream of the reheat coil. The submaster then manipulates its steam valve as necessary to provide the desired zone temperature.

The control system shown in Fig. 9.1 appears in Ref. (2) and is common to most reheat installations. Fresh air is mixed with return air to hold a 55-°F mixed-air temperature. In the heating season, the reheaters must supply enough steam to heat this 55-°F air to whatever temperatures are required

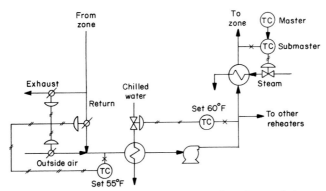

Fig. 9.1. A Multiple-zone reheat system with uncoordinated controls has no protection against simultaneous heating and cooling.

in the individual zones. Fresh air is not apportioned on a basis of need, but to satisfy an arbitrary specification of 55°F for mixed air. Thus fresh airflow is more than what is required as long as heat must be applied.

However in instances where solar incidence or internal heat generation cause a higher than desirable zone temperature, the controls will not allow a reduction in mixed-air temperature below 55°F.

When the mixed-air controller cannot keep its temperature below 60°F, the controller at the fan discharge introduces chilled water. This arrangement is particularly inefficient when outside-air temperature is in the 60–70-°F range, expecially when the humidity is high. Outside air is chilled and dehumidified, and then heated again to satisfy the master controllers. The dehumidification process represents a relatively high refrigeration load; yet even with reheat to 70°F, the air is uncomfortably cool for the time of year. Furthermore, an increase in the setting of the master controller will increase steam consumption, raise return-air temperature, and thereby also increase the cooling load. These uncoordinated controls waste energy while providing discomfort.

The system shown in Fig. 9.2 coordinates the control over heating, cooling, and ventilation to save energy and improve comfort. Similar to other systems appearing in this book, it is based on controlling the position of a selected valve.

The steam valves in these systems customarily fail open; they typically move from an open to a closed position as the air signal increases from 4 to 8 psig. If the signals to all reheater steam valves for a unit are compared in a *high* selector, the most-closed valve is then to be held closed at 9 psig by the valve-position controller, which manipulates fresh air and chilled water in sequence. If the selected valve signal is below 9 psig, the VPC will gradually close the chilled-water valve and then the fresh-air damper, until it returns

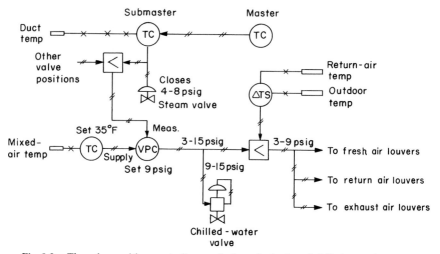

Fig. 9.2. The valve-position controller manipulates fresh air and chilled water in sequence, when all steam valves are closed.

to 9 psig. If heat is required in the steady state to satisfy the demands of the master, both the chilled water and fresh air will be off. Simultaneous heating and cooling is then avoided for the selected zone, although it may exist in an unselected zone. If a *low* selector is used, as shown in Fig 9.2, simultaneous heating and cooling is avoided for *all* zones, with temperatures rising above control points in the unselected zones.

   In many installations, one of the zones will consistently use more steam than the others, due to its location, occupancy, airflow pattern, or special needs. Then the signal to its steam valve can be connected directly to the VPC, since automatic selection is really superfluous. Where there is no zone which clearly leads the others in steam demand, one valve-position signal which is typically representative of the group may be used for control. This greatly simplifies the control system, especially where there are substantial distances between the locations of the steam valves. This is more likely to be encountered in the dual-duct systems described later, rather than in reheat systems, where all reheaters are grouped close to the fan.

   Coordination extends the operating range of the system. A high degree of heat release, or insolation that would cause zone temperature to rise, will shut off its steam valve. The VPC will then react to reduce mixed air temperature to as low as 35°F, if necessary, to bring the zone back to the verge of temperature control. The mixed-air controller limits temperature at 35°F by overriding the VPC, to protect the chilled-water coils from freezing.

   In the 60–70-°F high-humidity regime, cooling is not applied unless full fresh-air flow is reached, with no reheat. This raises zone humidity, increasing comfort without either heating or cooling. Should zone temperature begin to rise, only enough cooling is applied to keep it near set point, with the steam valve closed. As with other valve-position controllers, integral-action alone is desirable, set slowly enough to allow the master control loop to respond to manipulation of cooling.

When the outside-air temperature exceeds the return-air temperature, it can provide no further cooling. At this point, a differential temperature switch TS closes the fresh-air louvers. This transfer can also be based on enthalpy difference, if the humidity of the return air is substantially different from that of the fresh air. This switch must be *deactivated* whenever the chilled-water system is not functioning. In most of the northern areas, the refrigeration unit is disabled between October and May. Should the ambient temperature rise abnormally high during this period, the temperature switch could close the outside-air louvers, eliminating cooling altogether. This is certain to happen if the switching is based on outside-air temperature alone. If it is based on the differential between outside- and return-air temperatures, then the return air can only rise as high as outside-air temperature.

A minimum-fresh-air damper is shown in Fig. 9.1, to provide the necessary minimum ventilation; it is automatically closed when the fan is off.

2. Variable Air Volume

In the reheat system described above, airflow to the zones was held constant, and air temperature was adjusted for comfort. In a variable-air-volume (VAV) system, air temperature is held constant, and flow to each zone is varied, in accordance with demands from the master controller. While this arrangement is viable for cooling (although not optimum), it is poor practice for heating.

If airflow increases when the master temperature falls, the influx of warmer air can restore that temperature to its set point. But the same level of comfort will not be achieved. Consider a minimum-load condition of still air maintained at a dry-bulb temperature of 66.5°F, giving an effective temperature of 62°F as in Table 9.2. Should an increase in heat losses cause airflow to increase to 300 ft/min, the controlled dry-bulb of 66.5°F will produce an effective temperature of only 56.5°F.

While increased airflow promotes cooling, it would be more efficiently applied as an independent stage of cooling rather than as a means of manipulating heat flow to a cooling coil. At low cooling loads, for example, maintaining a constant air temperature of 60°F with a low flow will consume more energy than a temperature of 70°F and a higher flow. Every effort should be made to maximize temperatures throughout the refrigeration system, and maximizing airflow will help.

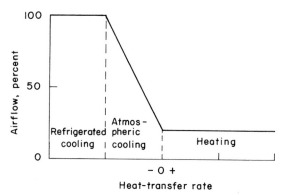

Fig. 9.3. Zone airflow should be minimum while heating, manipulated for control of atmospheric cooling, and held at maximum for refrigerated cooling.

Figure 9.3 outlines a recommended program for airflow control as a function of heating and cooling loads. As long as heating is necessary, zone airflow should be held at the minimum value consistent with reasonable distribution. When cooling is required, outside air should be used, with no increase in recirculating airflow, since this requires no additional energy. However, if zone airflow is controlled by throttling at the zone inlet, that damper should be opened for increased cooling.

When the outside-air damper is fully open and more cooling is required, the speed of the recirculation fan can then be increased to raise zone airflow. Only when full speed has been reached should chilled water be applied to cool the air.

This program can be implemented by combining the operation of heating and atmospheric cooling, as provided in the reheat system of Fig. 9.2. However, recirculation rate should be sequenced with chilled-water cooling, as both require energy. In essence, airflow would be increased to provide cooling up to its maximum, whereupon the chilled-water valve would begin to open. This arrangement is described in further detail in Section C,1 of this chapter.

Minimum ventilation requirements can be met in the heating regime by fixing a minimum outside-air damper position while the fan is operating at minimum speed. While atmospheric cooling is applied, outside airflow always exceeds the minimum. However when outside air enthalpy exceeds that of the return air during refrigerated cooling, outside air is returned to a minimum. But since the fan is then operating at full speed, a different position for the minimum outside-air damper is required.

Variable-volume systems employing both induced-draft and forced-draft fans require static pressure control within the building to minimize infiltration, particularly at high flow rates. Outside airflow may then have to be measured and controlled to ensure that a minimum is being maintained. This is more accurate than attempting to maintain a constant difference between the much larger supply and return flows.

3. Dual-Duct Systems

These systems provide two supplies of air—one warm and one cool—to each of the zones. Individual zone temperature controllers blend the necessary portions of each to satisfy the local demand. Blending has been described as an irreversible operation and can be the cause of wasted energy in these systems. Energy waste can be minimized by minimizing the temperature difference between the ducts.

The principal advantage of a dual-duct system is the ability to satisfy a variety of heating and cooling loads at individual zones. This can be accomplished most economically if the zone requiring the most heat or least cooling is drawing air fully from the warm duct, and the zone requiring the most cooling or least heating draws air only from the cool duct. All intermediate zones would then be blending, but the increase in entropy would be the minimum possible to satisfy these demands with the two-duct system.

Valve-position controllers are again recommended to set duct temperatures as shown in Fig. 9.4. Each zone-temperature controller drives a set of louvers adding warm air on a decreasing signal and cool air on an increasing signal. The lowest of the zone control signals will belong to the louver

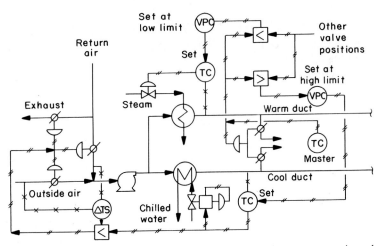

Fig. 9.4. Warm-duct temperature is set by the zone needing the most warm air, and cool-duct temperature by the zone needing the most cool air; chilled water is sequenced with atmospheric cooling.

drawing the most warm air. This lowest signal is controlled at the limit of travel by a VPC setting warm-duct temperature. A high signal selector sends the position of the louver drawing the most cool air to another VPC; this device attempts to hold that louver at the high limit of travel by setting cool-duct temperature.

Figure 9.4 shows the outside-air damper sequenced with chilled water so that refrigeration will not be used until atmospheric cooling is at a maximum. The presence of atmospheric cooling will increase the steam-flow requirements for the warm duct. Yet this arrangement is preferred over sequencing atmospheric cooling with the steam valve, which would require additional chilling. Since 1 Btu of low-pressure steam generates only 0.4–0.6 Btu of chilled water in the 40–50-°F range, more steam can be saved by reducing the cooling load than by reducing the heating load. In installations using independent fans for each duct, the warm-air duct should be supplied with only the minimum outside-air flow; this reduces steam requirements during periods of atmospheric cooling.

Although it is common practice to program warm-duct temperature as a function of outside-air temperature (4), it is not as capable of satisfying demands with minimal energy. If, at a particular outside-air temperature, warm-duct temperature is higher than necessary to meet zone loads, due to insolation, occupancy, or other factors, energy will be wasted. If it is insufficient, comfort requirements will not be met. The combination of selectors and valve-position controllers satisfies both requirements, under any load conditions within the capability of the HVAC unit.

4.   Air–Water Systems

In these systems, centrally conditioned air is distributed to individual zones where additional heating and/or cooling is provided by water. In the simplest arrangement, only hot water is available during the heating season, and only chilled water during the cooling season. Usually the same temperature controller and valve are used for both, requiring the action of the controller to be reversed at the time of the water changeover. There are special pneumatic controllers available whose action can be changed by raising the pressure of the air supply from 15 to 20 psig. An air switch on the main control panel selects the air supply for heating or cooling, at the time when the water-supply system is changed.

These systems have a limited operating range, and changeover between hot and chilled water is sufficiently troublesome that it is only performed twice a year. The operating range can be extended while conserving energy, by setting water temperature with a VPC acting on the most-open valve, and air temperature with a VPC acting on the most-closed valve as in Fig. 9.5. These valves typically fail open, so that the highest signal corresponds to the

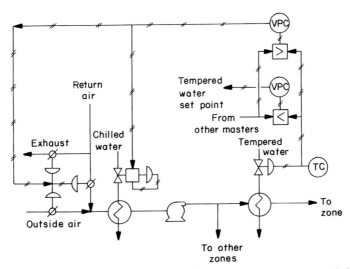

Fig. 9.5. The most-open valve position is used to set water temperature, and the most-closed valve position controls cooling by outside air and chilled water in sequence.

most-closed valve. Because the action of the zone-temperature controllers is reversed at changeover, the action of the valve-position controllers must also be reversed.

More elaborate systems have both hot and chilled water piped to the zone heat exchangers, with temperature controlled by a three-way mixing valve or sequenced two-way valves. Control over these systems is essentially the same as for the dual-duct system.

5.   Heat Pumps

There is no question that the space heating system of the future is the heat pump. There is no need at this point to develop the theory or application of the heat pump—it is adequately described in reference to refrigeration systems in Chapter 5. The work required to transfer a given amount of heat from a space into the surroundings is given as a function of temperature difference in Eq. (5.5). When heat is pumped into the space, it is augmented by the work applied. Because the work required is directly proportional to the product of heat transfer and temperature difference, and because heat flow in space conditioning is essentially proportional to temperature difference, W varies as $(\Delta T)^2$. Energy savings using heat pumps are then doubly sensitive to the thermostat setting. With a conventional heating system, reducing indoor dry-bulb temperature from 70 to 65°F against a 30-°F ambient would bring fuel savings of perhaps 14%; in a heat-pumped system, the same adjustment would provide a savings of nearly 31%.

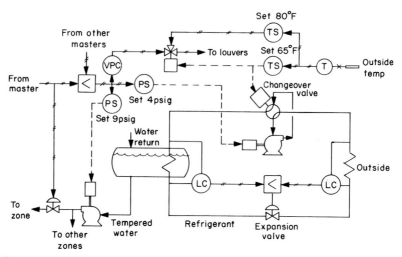

Fig. 9.6. With the changeover valve in its heating position as shown, the level controller on the water tank will drive its output upscale and be rejected by the low selector.

Heat pumps lose thermodynamic efficiency as ΔT increases, and are therefore not as attractive in northern climes. In addition, fouling of the outside heat-transfer surface with frost is a problem in cold, high-humidity environments. The advantage of being able to cool as well as heat makes the heat pump more attractive for moderate climates, rather than colder latitudes where summer cooling is unnecessary.

Controls for a heat pump are essentially no different than for any refrigeration unit. There are, however, the additional complications of changeover from heating to cooling, and a ventilation system to provide cooling when the weather is mild. Figure 9.6 shows the controls for a heat pump supplying tempered water to an air-conditioning system. In this illustration, a single master controller is shown. Assume it is capable of increasing tempered water flow on a temperature falling below 66°F, and also one rising above 80°F (such a controller is developed in Section C,1 following). When *all* zone tempered-water valves are closed, i.e., the lowest position signal is above 8 psig, the VPC in Fig. 9.6 will cool by admitting outside air. When one zone tempered-water valve reaches full opening at 4 psig, a pressure switch starts the compressor, which raises water temperature until that valve starts to close. A second pressure switch stops the water circulating pump when all water valves are closed.

An outdoor-air temperature sensor operates the changeover valve. When the ambient is below 65°F, the valve is in the heating position as shown. Refrigerant vapor is directed downward into the water coil, where it condenses. The level controller is intended to hold the coil full of condensate when the flow is in the other direction. Consequently, it will attempt to open the expansion valve to raise the level, but only succeeds in lowering it. The level controller in the evaporating coil will respond correctly to a rising level by throttling the expansion valve, and will therefore be selected for control.

When the outdoor temperature reaches 65°F, a temperature switch changes the compressor flow to a cooling cycle and locks the fresh-air louvers in an open position. When any zone control signal is then below 9 psig, the water circulating pump is started, and the fresh-air louvers remain open. When the first zone control signal reaches 4 psig, the compressor will start, to cool the tempering water. When the outdoor air reaches 80°F, the fresh-air louvers automatically close. (This last action could also be performed on the basis of temperature or enthalpy difference between outside and return air as described earlier.)

Driving a heat pump with a diesel engine can add to the efficiency of the system. Along with other advantages already considered, the heat rejected by the engine and its exhaust may be added to that pumped by the compressor. During the cooling season, the exhaust heat must be rejected, but engine heat may still be used to provide hot tapwater.

An engine has the advantage of variable-speed operation, as opposed to the electric-motor driven compressor described earlier, which was either on or off. The control system can be essentially the same as that shown in Fig. 9.6, except for the addition of a valve-position controller set at 4 psig, to adjust the engine's governor.

Figure 9.7 shows how the engine temperature can be controlled while providing hot tapwater. Heat in excess of that required for tapwater is sent

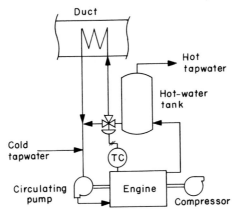

Fig. 9.7. Engine heat may provide hot tapwater and supplement space heating in winter.

to the air duct to supplement pumped heat in winter. During the cooling season, this duct must be diverted to outside air by the changeover controls. The tapwater will require supplementary heat during mild conditions.

C. HVAC MANAGEMENT

Section B examined the characteristics of various HVAC systems and applied controls to them to maximize their efficiency. Simultaneous heating and cooling was minimized, and outside air was used as much as possible. What remains is to manage these HVAC systems to provide the comfort conditions found to be most conservative in Section A. One of the most important techniques to be developed is the programming of heating and cooling to minimize the use of either, over as wide a range of conditions as possible. Another energy-saving feature to be considered is the programming of heating and cooling as a function of occupancy and activity. Humidity control for special environments is also included.

1. Maintaining Effective Temperature

The most common complaint about air-conditioning systems in general is their failure to maintain comfortable conditions while consuming excessive quantities of energy. This is due in large measure to control of dry-bulb temperature at a single point year round, irrespective of humidity and air-velocity, as described earlier. Furthermore, there has been little, if any, consideration of acclimatization by the metabolism. Typically, controlled conditions are cooler than necessary in summer and warmer than necessary in winter, except during extreme loads when control is lost altogether. In short, energy has been used to cause discomfort.

Master control points must be different for heating and cooling to take advantage of the wide range of conditions over which neither is necessary. If two controllers are used, as shown in Fig. 9.8, control points are independently adjustable as a function of season and preference. This system is intended to drive a tempered-water valve as would be used with the heat pump system of Fig. 9.6.

The heating controller would have high-gain proportional or proportional-plus-integral action to minimize the heat applied above the set point. The valve-position controller of Fig. 9.6 will increase ventilation to control zone temperature only when the hot water is off to *all* zones. Therefore the coolest zone would be controlled at set point by ventilation, while all other zones would be warmer. The zone temperature at which atmospheric cooling is applied could be increased by eliminating integral action from the master controller. However, then some heating will continue to be applied when

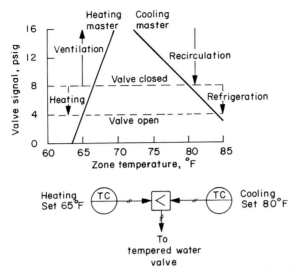

Fig. 9.8. When manipulating a tempered-water valve, the heating and cooling controllers must have opposite actions.

zone temperature is above set point. It should be remembered that atmospheric cooling is free—there is no heating, refrigeration, nor increase in fan power required to provide it.

When the zone temperature approaches the set point of 80°F, the fan speed should be increased to maximum, as described under variable-air-volume controls. At 80°F, the tempered-water valve begins to open again through the action of the low-signal selector. Proportional action in this range should program the valve to open fully at 84°F. The reasoning behind this program can be explained in reference to Table 9.6. An effective temperature of 73°F can be attained at about 80°F dry-bulb, when the dew point of the air is 73°F. This high dew point will be achieved when the cooling load is minimal, if the chilled or tempered-water control system adjusts water temperature as a function of cooling load. This will be automatically achieved by the heat-pump controls in Fig. 9.6, and by the chilled-water controls described in Figs. 5.19–5.22. Then as the heat load calls for increased water flow, water temperature will be adjusted downward, reducing the dew point of the air, thereby allowing an increase in dry-bulb temperature up to 84°F.

The speed of the recirculating fan can be driven directly by the lowest output of the master controllers, as shown in Fig. 9.9. When all chilled or tempered-water valves are closed, the selected master controller will adjust fan speed over the 8–15-psig range. In heat-pump systems, this action must be inhibited when heating, so that minimum speed will be provided. This logic can be initiated by the changeover controls.

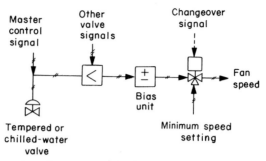

Fig. 9.9.   Fan speed may be programmed to follow the master controller requiring the most cooling.

When applied to dual-duct or air-water systems having both hot and cold supplies, both heating and cooling masters must have the same action. Selection between the two masters must then be established by the demand for increased circulation or for chilled water. When cooling by ventilation is no longer sufficient to maintain comfort at the setting of the selected heating master in the system of Fig. 9.4, the chilled-water valve will begin to open. A pressure switch acting on the signal to the chilled-water valve can be used to switch all masters on that unit to the cooling controller. Because the cooling controller set points are higher, the chilled-water valve will immediately close, deactivating the pressure switch. Cycling about this point will take place, unless a delay is provided. The system in Fig. 9.10 shows a time-

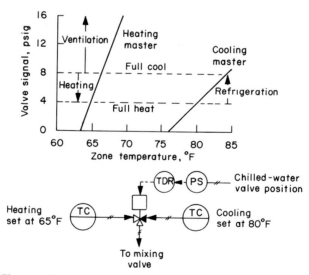

Fig. 9.10.   The controller with the higher set point is used whenever chilled water is required.

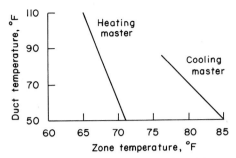

Fig. 9.11. In systems having a submaster controller for each master, duct temperature should be coordinated with zone temperature as shown.

delay relay holding in the cooling masters for a set interval following the initial opening of the chilled-water valve.

With reheat systems using a submaster duct controller with each master, the two temperatures can be coordinated as shown in Fig. 9.11. Two proportional controllers are required, with the same action but with different gains and set points; selection between them should be based on chilled-water valve position, as in Fig. 9.10.

2. Seasonal Programming

The programs for master temperature controllers shown in Figs. 9.8–9.11 show the extreme settings for summer and winter operation. Considering that acclimatization takes time, it would be wise to lower the cooling controller set point somewhat in winter, and to raise the heating controller set point somewhat in summer.

Ideally, these settings ought to be adjusted several times a year. This would be prohibitive for multiple-zone systems where all master controllers were located in their respective zones. However if only transmitters were located in the zones, with all master controllers housed in a central facility, set-point adjustments could be made to all controllers from a single station. A program such as that shown in Table 9.7 would seem reasonable.

Table 9.7

Master Set Points at Various Times of Year

Months	Heating set point (°F)	Cooling set point (°F)
Dec–Mar	65	75
Apr–June	68	78
July–Sept	72	82
Oct–Dec	68	78

3. Diurnal Programming

Substantial quantities of energy can be saved by programming set points between night and day, or periods of occupancy. In an office or manufacturing building, the heating set point can be reduced to 55°F after the occupants have left for the day, if a separately controlled space is available for resident security personnel. At the same time, ventilation and recirculation can be reduced to zero. These effects are simultaneously achieved by controlling the recirculating fans with a night thermostat. A 7-day timer switches control of each fan to a single thermostat set at 55°F, at night and on weekends. The thermostat stops the fan until the temperature falls to that set point, whereupon it is started. Because heat losses are low in the absence of recirculation, the fans stay off for long periods during the unoccupied hours.

This 10-°F reduction in temperature will result in a 40% reduction in indoor–outdoor temperature difference, when outdoor air is at 30°F. In the absence of recirculation, the energy savings should exceed 40%.

During the cooling season, fans may be turned off without any control, when the building is unoccupied, at least at night. Reduction in lighting and absence of people, as well as lack of insolation, will keep temperatures from climbing.

4. Humidity Control

Certain buildings or areas of buildings are used to store or work with materials sensitive to relative humidity. The storage and printing of instrument charts, films, and graph paper fall into this category. Close control over dry-bulb temperature and humidity must be maintained at a single set of conditions year round.

Control of temperature can be simplified and rendered insensitive to ambient conditions if simultaneous heating and cooling are allowed. This common practice is often combined with simultaneous humidification and dehumidification to achieve humidity control. Yet it is possible to provide satisfactory control over both temperature and humidity with a minimum of energy by properly coordinating controls as in Fig. 9.12.

The figure shows heating using steam, and humidification with a water spray. Refrigeration, providing both cooling and dehumidification, is operated on an on–off basis. When heating and humidification are required, both valves are open, and the pressure switches keep the refrigeration unit off. As the heat load decreases, a VPC on the steam valve begins cooling by ventilation. Should the steam valve be driven to a fully closed position, a pressure switch will start the refrigeration unit.

Because of the stepwise nature of the applied refrigeration, a small amount of steam will begin to flow to maintain temperature control. When the dif-

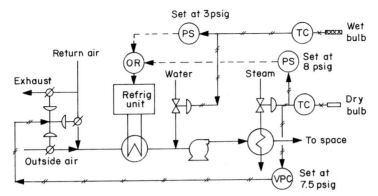

Fig. 9.12.   The refrigeration unit is operated whenever the steam valve *or* the water valve is closed.

ferential gap in the pressure switch is exceeded, the refrigeration unit will be stopped, and the steam valve will again begin to close.

Should the water valve close fully, indicating the impending loss of humidity control, its pressure switch will start the refrigeration unit, even with the steam valve open. A differential gap is necessary in this pressure switch as well, to extend the cycle time of the refrigeration unit.

Note that wet-bulb temperature is controlled rather than humidity. At a controlled dry-bulb temperature, a constant wet-bulb temperature will provide a constant humidity. The advantage of using wet-bulb temperature is its insensitivity to water injection. When room-temperature water is evaporated into room-temperature air, the dry-bulb temperature will fall, although the wet-bulb temperature is unaffected. This eliminates the interaction that normally exists between temperature and humidity control loops. Adding heat to the air raises both dry- and wet-bulb temperatures, but that disturbance cannot be propagated by the water valve back to the wet-bulb temperature controller.

D.   SOLAR ENERGY SYSTEMS

The most useful application of solar energy will be the heating of buildings and tapwater. While the heating medium is free, it is also quite variable, necessitating storage and auxiliary sources of energy for complete service. Proper control can minimize the use of these auxiliaries.

1.   Flat-Plate Collectors

The solar-energy collector most useful for heating buildings and tapwater is the fixed, flat-plate unit. It consists of a radiant-energy absorbing surface positioned to receive as much sunlight as possible during the course of the day. In northern latitudes, the collector should face due south, at an inclination from the horizontal equal to the latitude of the location. To maximize absorption seasonally, the inclination should be increased 15 deg in winter and decreased 15 deg in summer, from that spring-fall position.

The heat-absorbing surface is insulated from the outside air typically by two translucent cover plates with air spaces between. The covers pass visible light but are opaque to infrared radiation, to minimize the reradiation of heat from the collector. The heat-absorbing surface may be cooled by the passage of air or water over, through, or behind the surface. Thermal insulation covers the dark side of the collector.

2.   Heat Transfer and Storage

Figure 9.13 shows a flat-plate collector used to provide hot air for space heating. A bed of crushed and graded stone is used as a heat storage element. Two differential-temperature switches are shown to maximize heat recovery. When the collector is warmer than the stone bed, both on–off dampers are open, allowing heat to flow into the bed. When the temperature difference reverses, flow into the stone bed is stopped by closing the damper at its top. Flow may continue through the collector into the space to be heated as long as the collector is warmer than the circulating air. When this temperature difference reverses, the damper in the stream leaving the collector is closed, its fan stopped, and the damper at the top of the stone bed is reopened. Then heat is removed from storage.

The duct temperature controller is shown manipulating heat and atmospheric cooling sequentially. Should the controller output reach full scale

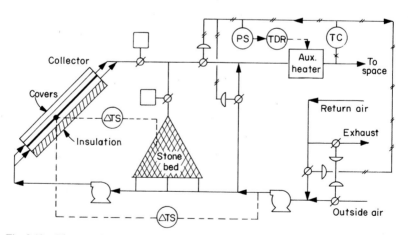

Fig. 9.13.   The stone bed is charged when it is cooler than the collector, and discharged when the collector is cooler than the stone bed and the circulating air.

for a sustained period, indicating insufficient solar heat to maintain temperature control, an auxiliary heater is started by a pressure switch acting through a time-delay relay. In multiple-zone systems of the variable-volume type, duct temperature could be set by the position of the most-open damper, through a valve-position controller. Energy-conservation techniques such as this are essential to the economic success of such low-grade heat sources as solar thermal energy.

Circulating water systems require protection against freezing. Antifreeze additives may be used, but their cost becomes prohibitive to protect the entire storage system, which may include thousands of gallons. Alternately, antifreeze protection may be applied only to the water circulating through the collector, with a heat exchanger located in the storage vessel to isolate it from the stored water. The heat exchanger adds significantly to the capital cost of the system, and also reduces its efficiency. Duffie and Beckman (5) estimate that each degree Celsius temperature drop across the exchanger reduces the collector efficiency by 1–2%. (Higher collector temperatures reduce absorption and increase heat losses.)

Another method to avoid freezing is to elevate the collector above the storage vessel so that it may drain freely when the circulating pump is stopped. This practice requires more head of the pump, but pumping energy is such a small fraction of the heat transfer through the system that the added cost is not significant.

Figure 9.14 shows how a solar heating system using water might be controlled. When the collector is warmer than the storage tank, water will

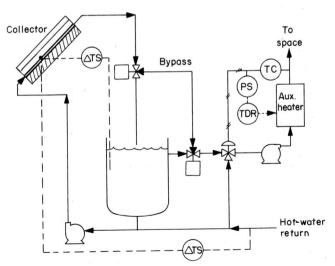

Fig. 9.14. The storage tank is bypassed when it is warmer than the collector, until the collector is cooler than the return water.

circulate through the two. When collector temperature falls below storage temperature, the tank is bypassed. When it finally falls below return-water temperature, no more heat absorption is taking place; then the collector circulator is stopped, and the bypass valves are returned to their normal position. The collector then drains, and heat is removed from storage only. When the sun rises again, and collector temperature increases above storage temperature, circulation through the collector will resume. Auxiliary heat is provided in the same manner as with the air systems.

3. Adding a Heat Pump

The storage capability in a solar heating system can be extended by means of a heat pump. When storage temperature approaches room temperature, control will ordinarily be lost. But at that point, a refrigeration compressor can be started to pump heat from the storage vessel into a heat exchanger in the water supply line. Figure 9.15 shows a pressure switch on the output of the water temperature controller activating the compressor. As the temperature in the storage vessel begins to fall below that of the heat exchanger, a differential temperature switch stops circulation through the storage tank.

Solar heating systems typically can increase storage temperature to about 140°F during the day. As heat is drawn from storage during the night, that temperature might fall to 100°F, depending on the weather, the size of the storage vessel, room temperature settings, etc. To size a tank for more than one or two day's capacity may not be economical. By adding a heat pump, however, storage temperature may be reduced to 32°F, virtually doubling the heat capacity of the system.

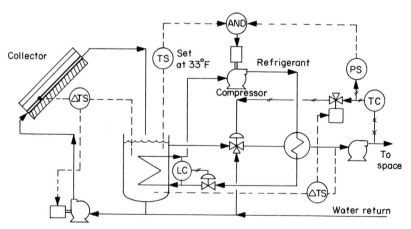

Fig. 9.15. The compressor is started whenever temperature control can no longer be attained by drawing from storage.

Because the power required of the heat pump varies with ΔT, comparatively little energy is used when the storage tank is near room temperature. Since operating under these conditions is much more probable than operation close to freezing, the heat pump generally runs near maximum efficiency and therefore at little cost.

To maximize the performance of this system, all of the techniques described earlier for saving energy should be applied. Thermostats should be at minimum allowable temperatures day and night to minimize heat-pump power and the likelihood of dropping storage temperature to 32°F, where no more heat would be available. Hot-water temperature should be set at the minimum necessary to satisfy room temperature settings. In a multizone system, this would be accomplished by a valve-position controller responding to the most-open water valve.

This system probably offers the best economy of all for northern climates. The only improvement might be the use of a diesel-engine drive, with recovery of engine and exhaust heat. The heat pump can also be used for air-conditioning, and the large water-storage vessel can effectively smooth day-night heat flux variations.

REFERENCES

1. "ASHRAE Handbook of Fundamentals," Chapter 7. American Society of Heating, Refrigeration, and Air-Conditioning Engineers, New York, 1972.
2. "Heating, Ventilating and Air-Conditioning Guide," Chapter 1. American Society of Heating and Air-Conditioning Engineers, New York, 1959.
3. "Energy Conservation Design Guidelines for Office Buildings," pp. 9–25. General Services Administration, Washington, D.C., 1974.
4. "ASHRAE Handbook and Product Directory—Systems," p. 3.33. American Society of Heating, Refrigeration, and Air-Conditioning Engineers, New York, 1976.
5. Duffie, J. A., and W. A. Beckman, "Solar Energy Thermal Processes," p. 267. Wiley (Interscience), New York, 1974.

Index